Regular Sequences and Resultants

Research Notes in Mathematics

Volume 8

Regular Sequences and Resultants

Günter Scheja
Universität Tübingen
Tübingen, Germany

Uwe Storch
Ruhr-Universität Bochum
Bochum, Germany

CRC Press
Taylor & Francis Group
Boca Raton London New York

CRC Press is an imprint of the
Taylor & Francis Group, an **informa** business

AN A K PETERS BOOK

First published 2011 by A K Peters, Ltd.

Published 2019 by CRC Press
Taylor & Francis Group
6000 Broken Sound Parkway NW, Suite 300
Boca Raton, FL 33487-2742

© 2001 by Taylor & Francis Group, LLC
CRC Press is an imprint of Taylor & Francis Group, an Informa business

First issued in paperback 2019

No claim to original U.S. Government works

ISBN 13: 978-0-367-45528-6 (pbk)
ISBN 13: 978-1-56881-151-2 (hbk)

Visit the Taylor & Francis Web site at
http://www.taylorandfrancis.com

and the CRC Press Web site at
http://www.crcpress.com

Library of Congress Cataloging-in-Publication Data

Scheja, Günter, 1932-
    Regular sequences and resultants / Günter Scheja, Uwe Storch.
        p. cm. -- (Research notes in mathematics ; v. 8)
    Includes bibliographical references and index.
    ISBN 1-56881-151-9
    1. Elimination. 2. Projective spaces. 3. Intersection theory. 4. Sequences (Mathematics)
    I. Storch, Uwe. II. Title. III. Research notes in mathematics (Boston, Mass.) ; 8.

QA192 .S32 2001
512.9'434--dc21   2001022483

Visit the Taylor & Francis Web site at
http://www.taylorandfrancis.com

and the Psychology Press Web site at
http://www.psypress.com

Preface

In these notes we develop aspects of elimination by treating regular sequences and resultants.

We consider quite generally anisotropic projective spaces which gained increasing attention in the last decades. Arbitrary noetherian rings will be used as base rings, and the representation runs throughout in terms of commutative algebra, which has the positive side effect that the pattern of specialization appears fitting. The key concepts are regular sequences and complete intersections. Their use benefits by the well-developed and clear duality theory.

A reading of these notes thus calls for a basic knowledge of commutative algebra and of homological methods, the features of Koszul complexes, for example. We made some effort, though, to give complete and down-to-earth proofs for those parts which depend on special constructions.

More details are supplied in the following descriptions of the individual chapters.

Chapter I deals with two topics which do not depend on the material of the other chapters and hence can be read independently.

Sections 1 and 2 treat Kronecker's method of indeterminates. The concept of Kronecker extensions of a ring and its modules is introduced and developed in a general set-up. These sections also provide typical examples of the kind of methods of commutative algebra used in the notes.

The second topic is the study of numerical monoids, i.e. submonoids of the additive monoid $\mathbb{N}$ of non-negative integers containing almost all elements of $\mathbb{N}$. Such a monoid turns up in a natural way as the monoid $\mathrm{Mon}(\gamma) = \mathrm{Mon}(\gamma_0, \ldots, \gamma_n)$ generated by the (positive) weights $\gamma_0, \ldots, \gamma_n$ of the indeterminates $T_0, \ldots, T_n$ in a polynomial algebra $A[T] = A[T_0, \ldots, T_n]$ which defines the anisotropic projective space $\mathbb{P}_\gamma(A) = \mathrm{Proj}\, A[T]$ over the commutative ring A. At first reading it suffices to look through Section 3.

Chapter II contains the treatment of two of the fundamental concepts which we will use extensively in the later chapters, namely the concept of regular sequences and the concepts of (relative) complete intersections and locally complete intersections. This chapter may also serve as an independent introduction into methods which have useful applications in algebraic geometry as well as in complex-analytic geometry.

Basic constructions and lemmata are treated in Section 6, whereas Section 7 features specialities of the graded case.

Crucial for what follows is the treatment of – as we call them – generic polynomials in Section 8. Generic polynomials are (not necessarily homogeneous) polynomials $\sum_{\nu \in N} U_\nu T^\nu \in Q[T] = Q[T_0, \ldots, T_n]$, where the set N of tuples of exponents $\nu = (\nu_0, \ldots, \nu_n)$ is a finite subset of $\mathbb{N}^{n+1} \setminus \{0\}$ and where the U_ν are (different) indeterminates in some polynomial ring Q over $\mathbb{Z}$. The fact that a sequence $F_0, \ldots, F_r$ of such polynomials is regular, and that the Lasker-Noether decomposition of the ideal they generate has in addition suitable properties, can be characterized by combinatorial means. This leads to the concepts of admissible and strictly admissible sequences of generic polynomials, which may well be of interest in other parts of algebra. As a special example we treat sequences of generic binomials.

In Section 9 we prove a combinatorial characterization of algebraic independence of generic Laurent polynomials in $Q[T, T^{-1}]$. This will be used frequently in what follows.

Chapter III presents the main case of elimination (with respect to projective spaces). In Section 10 the necessary prerequisites are briefly developed. The elimination ideal $\mathfrak{T}_0$, part of the ideal of inertia forms $\mathfrak{T}$, is introduced, and the so-called main theorem of elimination is proved.

In Section 11 we switch to the main case of elimination, which centres around the case of $n + 1$ homogeneous polynomials $F_0, \ldots, F_n$ in $n + 1$ indeterminates $T_0, \ldots, T_n$. In a first step generic sequences are treated. For many purposes it suffices to handle regular sequences of generic polynomials, i.e. admissible ones. In the strictly admissible case more satisfactory results are derived, including the computation of the degree of elimination.

Section 12 is a substantial part of the notes. The ideal of inertia forms and the elimination ideal with respect to an arbitrary regular sequence of homogeneous polynomials $F_0, \ldots, F_n \in A[T_0, \ldots, T_n]$ are determined. The means for this are provided by an extended version of duality for graded complete intersections, which is developed and discussed in detail. Thereby the ideal of elimination is shown to be divisorial if A is an integrally closed noetherian domain, hence a principal ideal $\neq 0$ if A is even factorial.

Chapter IV, finally, deals with the resultants. To begin with we introduce in Section 13 the resultant ideal $\mathfrak{R} = \mathfrak{R}(F_0, \ldots, F_n)$ for a regular sequence $F_0, \ldots, F_n$ of homogeneous polynomials of positive degrees in the polynomial algebra $A[T_0, \ldots, T_n]$ over an integrally closed noetherian domain A. The ideal $\mathfrak{R}$ in A is a divisorial ideal like the elimination ideal $\mathfrak{T}_0$ of the polynomials and has the same zero set as $\mathfrak{T}_0$, but it has better functorial properties and is always principal (which is not true for $\mathfrak{T}_0$ in general). Also it suits convincingly the geometric situation: The degree of $\mathfrak{R}$ defines the degree of elimination in a proper way. This is the main result of Section 14 and is again an application of straight duality theory for regular sequences. In addition, these methods yield further characterizations of the resultant divisor.

In Section 15 we use the Koszul resolution to construct a canonical generator $R = \mathrm{R}(F_0, \ldots, F_n)$ of the resultant ideal $\mathfrak{R}(F_0, \ldots, F_n)$, which we call the resultant. Going back to the generic situation one can define the resultant for an arbitrary sequence $F_0, \ldots, F_n$ of homogeneous polynomials in $A[T_0, \ldots, T_n]$ over any base ring A in a unique way up to sign. In the classical case with weights $\gamma_0 = \cdots = \gamma_n = 1$ the element $R \in A$ is the classical resultant already known in the 19th century and described, for instance, by A. Hurwitz in [19]. In the last section we present some properties of resultants which are valuable for calculations, for example the norm formula and the product formula.

Supplements are added to every section. Some simply provide exercises fitting the context or supply further information and examples, for instance showing that inequalities are sharp or that certain hypotheses are indispensable. Some describe details of tools or results generally known in commutative algebra, but not readily available in the literature. And some add material to special points we like to emphasize but couldn't pursue in the main lines of development.

Added hints to crucial arguments and other remarks, put in brackets, are sometimes elaborate, at least in those supplements serving the main text directly, where the reader should not be left alone. The comments are also detailed in supplements which follow a theme of their own and in which the reader might be interested for various reasons.

In general, we have taken care that it is unnecessary for the reader to go through the supplements in order to be able to follow the main ideas. In addition, those supplements deepening sidelines or following historical paths or those which may finally be of interest to specialists only, are given the sign † at the beginning as a warning. These supplements can safely be skipped in a first reading.

Prerequisites on commutative algebra are best taken from Atiyah/Macdonald [1], Serre [45], Matsumura [31] and then Bourbaki [4]. Simple homological methods or advice on how to find convenient sources of them can be taken from Matsumura [31], Bruns/Herzog [6], Eisenbud [12] and Hilton/Stammbach [18]. Only few concepts of algebraic geometry are being used, especially however the concept of the projective spectrum. For basic structures of this kind we refer to Hartshorne [15].

Discourses on resultants have been abundant for a long time. A historical picture can be gathered from the papers [21], [22], [23], [24], [25] of Jouanolou, where many extensions are developed in terms of modern algebraic geometry.

Characteristics of elimination from a more geometric standpoint and a report on the corresponding literature are provided by Gel'fand/Kapranov/ Zelevinsky [14]. The text book Cox/Little/O'Shea [9] can be recommended for a first reading.

Our interest in the subject of regular sequences and resultants started in 1988 when there seemed to be no modern account of the classical treatment as, say, that given in 1913 by Adolph Hurwitz [19]. So we made up our own version of the subject starting from Hurwitz's article and generalizing to arbitrary noetherian commutative ground rings. Some notes on the subject which we have written down occasionally over the years, whenever we found time for it, are now put together for these notes. We could not deal with all of the developments of elimination theory in recent years without unduly enlarging the text. Instead we restricted our attention to those parts of the theory which are closely related to the concept of complete intersection, influenced by [41], [42], [28] and later by [43], [44].

During the preparation of these notes we were aided by many people. We are grateful to David Eisenbud and the Fellowship of the Ring at Brandeis University, Mass., for profitable discussions and general assistance during our stay there in February/March 1992. It was supported by the Deutsche Forschungsgemeinschaft (DFG) which we gratefully acknowledge. On several occasions the second author had the opportunity to report on the progress of these notes in a series of lectures at the Indian Institute of Science at Bangalore organized by Dilip Patil and supported by the Deutscher Akademischer Austauschdienst (DAAD).

Many helped us by pointing out details of interest, putting questions to us or providing us with background material. In particular, we mention Gerd Müller, Jürgen Herzog, José Gómez Torrecillas, Dilip Patil and our students Goran Dević, Benjamin Nill as well as people in our lectures and talks on the subject. We would like to express our sincere thanks to them.

Last but not least, we cordially thank Wilhelm Kaup and Klaus Peters for their help in editing these notes.

Bochum and Tübingen   G. Scheja and U. Storch

Contents

I Preliminaries

    1   Kronecker Extensions   1

    2   Modules and Kronecker Extensions   6

    3   Numerical Monoids   13

    4   Relations of Numerical Monoids   23

    5   Splitting of Numerical Monoids   36

II Regular Sequences

    6   Regular Sequences and Complete Intersections   41

    7   Graded Complete Intersections   53

    8   Generic Regular Sequences   60

    9   The Generic Structure of the Principal Component   75

III Elimination

    10   Basics of Elimination   81

    11   The Main Case for Generic Regular Sequences   85

    12   The Main Case for Regular Sequences   96

IV Resultants

    13   Resultant Ideals   105

    14   Resultant Divisors and Duality   109

    15   Resultants   119

    16   Formulas on Resultants   129

References   137

Index   141

I Preliminaries

1 Kronecker Extensions

In this section and in the next one we define Kronecker extensions and develop their basic properties in terms of commutative algebra. Kronecker extensions provide the conceptual tools for Kronecker's *method of indeterminates* ("Unbestimmten-Methode"). Special cases of it have been used for a long time, cf., for instance, Supplement 7 in Section 2. A modern use of the method can be found in Nagata's book [33].

In the following A always denotes a commutative ring.[1] $\operatorname{Spec} A$ is the prime spectrum of A and $\operatorname{Spm} A$ the maximal spectrum of A, both of them endowed with the Zariski topology.

The content $C(F)$ of a polynomial F over A (in any set of indeterminates) is the ideal in A generated by its coefficients. F is called primitive if its content is the unit ideal.

For a system U_i, $i \in I$, of indeterminates over A the short notation

$$A[U] := A[U_i : i \in I]$$

will be used whenever this is not subject to misinterpretation.

A polynomial $F \in A[U]$ is primitive if and only if for any maximal ideal $\mathfrak{m}$ in A the image of F in $(A/\mathfrak{m})[U]$ is not the zero polynomial. In other words: The complement of the set S of all primitive polynomials in $A[U]$ is the union of all the extended ideals $\mathfrak{m}A[U]$, $\mathfrak{m} \in \operatorname{Spm} A$.

1.1 Lemma *The set S of primitive polynomials in $A[U]$ is a saturated multiplicative system of non-zero-divisors.*

Proof. By the last remark above, S is a saturated multiplicative system. The fact that a primitive polynomial is a non-zero-divisor simply follows from the well-known Lemma of McCoy which states: *If $F \in A[U]$ is a zero-divisor then there is an element $a \in A$, $a \neq 0$, with $aF = 0$.* □

We now introduce Kronecker extensions.

Definition For an arbitrary family U_i, $i \in I$, of indeterminates the A-algebra

$$A(U) = A(U : i \in I) := A[U_i : i \in I]_S ,$$

---

[1] A commutative ring is a commutative ring with identity. All modules are unitary.

that is the ring of fractions of $A[U]$ with respect to the multiplicative system S of primitive polynomials, will be called the **Kronecker extension** of A (in the indeterminates U_i, $i \in I$).

1.2 Proposition *Every Kronecker extension $A \to A(U)$ is faithfully flat.*

Proof. $A \to A(U)$ is the composition $A \to A[U] \to A[U]_S$ and therefore flat. Because of $\mathfrak{m}A[U] \cap S = \emptyset$ for $\mathfrak{m} \in \operatorname{Spm} A$ we have $\mathfrak{m}A(U) \neq A(U)$. Thus the extension is even faithfully flat.   □

Especially $A(U)^\times \cap A = A^\times$.

1.3 Proposition *The canonical mapping $\operatorname{Spec} A(U) \to \operatorname{Spec} A$ induces a homeomorphism $\operatorname{Spm} A(U) \to \operatorname{Spm} A$. Every maximal ideal of $A(U)$ is the extension of a maximal ideal of A.*

Proof. $\operatorname{Spm} A(U)$ can be identified with the subspace M of $\operatorname{Spec} A[U]$ consisting of those (prime) ideals which are maximal in the set of all ideals of $A[U]$ having empty intersection with S.

For $\mathfrak{P} \in M$ the following is true: A polynomial $F \in A[U]$ belongs to $\mathfrak{P}$ if and only if all its coefficients belong to $\mathfrak{P}$, i.e. $C(F) \subseteq \mathfrak{P}$. This is trivial for $I = \emptyset$, because then simply $A(U) = A$. Assume therefore $i \in I$. Let a be a coefficient of $F \in \mathfrak{P}$ such that $a \notin \mathfrak{P}$. Then there is a primitive polynomial of the form $G + aH$, $G \in \mathfrak{P}$, $H \in A[U]$, and the ideal $C(G) + aA$ is the unit ideal. On the other hand, for $n > \deg F$ we have $F + U_i^n G \in \mathfrak{P}$ and $A \neq C(F + U_i^n G) = C(F) + C(G) \supseteq aA + C(G)$, a contradiction.

As a consequence every $\mathfrak{P} \in M$ is an extended ideal:

$$\mathfrak{P} = (\mathfrak{P} \cap A)A[U]$$

where $\mathfrak{P} \cap A$ is clearly maximal in A. Conversely, $\mathfrak{m}A[U] \in M$ for every $\mathfrak{m} \in \operatorname{Spm} A$. Furthermore, for any ideal $\mathfrak{A}$ in $A[U]$,

$$V(\mathfrak{A}) \cap M = V(\mathfrak{a}A[U]) \cap M$$

where $\mathfrak{a}$ is the ideal generated in A by the coefficients of all elements of $\mathfrak{A}$. Thus the projection $\operatorname{Spm} A(U) \to \operatorname{Spm} A$ is not only bijective but a homeomorphism, too.   □

As a corollary to 1.2 and 1.3 we have that *Kronecker extensions preserve the length of modules.*

1.4 Concordance rule *Let $A \to A'$ be an integral homomorphism of rings. Then*

$$A'(U) = A' \otimes_A A(U).$$

Proof. A homomorphism $A \to A'$ induces a homomorphism $A(U) \to A'(U)$ and hence a canonical homomorphism $A' \otimes_A A(U) \to A'(U)$, which is simply the inclusion

$$A' \otimes_A A[U]_S \;=\; A'[U]_T \;\subseteq\; A'[U]_{S'}$$

where T is the canonical image of S in $A'[U]$ and S' is the set of all primitive polynomials of $A'[U]$. We have to prove that S' is the saturated hull of T (assuming $A \to A'$ being integral).

Let $\mathfrak{P}'$ be any prime ideal in $A'[U]$ with $\mathfrak{P}' \cap T = \emptyset$. We have to show then that also $\mathfrak{P}' \cap S' = \emptyset$.

For $\mathfrak{P} := \mathfrak{P}' \cap A[U]$ [1] we have $\mathfrak{P} \cap S = \emptyset$. Thus, by 1.3, there is a maximal ideal $\mathfrak{m}$ in A such that $\mathfrak{P} \subseteq \mathfrak{m}A[U]$. By the going-up theorem there is a prime ideal $\mathfrak{Q}'$ in $A'[U]$ such that $\mathfrak{P}' \subseteq \mathfrak{Q}'$ and $\mathfrak{Q}' \cap A[U] = \mathfrak{m}A[U]$. Furthermore, $\mathfrak{Q}'$ is a minimal prime of $\mathfrak{m}A'[U]$. Again using the assumption that $A \to A'$ is integral we see that the prime ideal $\mathfrak{m}' := \mathfrak{Q}' \cap A' \supseteq \mathfrak{m}A'$ is maximal in A'. Because $\mathfrak{Q}'$ is minimal over $\mathfrak{m}'A'[U] \supseteq \mathfrak{m}A'[U]$ we have $\mathfrak{Q}' = \mathfrak{m}'A'[U]$ and thus $\mathfrak{Q}' \cap S' = \emptyset$, from which $\mathfrak{P}' \cap S' = \emptyset$ follows. $\square$

For the case of a residue class ring $A' = A/\mathfrak{a}$ there is a simple direct proof of 1.4.

If A is local with maximal ideal $\mathfrak{m}$ and residue field $k = A/\mathfrak{m}$, Propositions 1.3 and 1.4 show that a Kronecker extension $A(U)$ is local with maximal ideal $\mathfrak{m}A(U)$ and the rational function field $k(U)$ as residue field. This construction was used extensively by Nagata in [33] to guarantee infinite residue fields.

1.5 Lemma *Assume A to be noetherian. Then for every $\mathfrak{p} \in \operatorname{Spec} A$:*

$$\operatorname{codim} \mathfrak{p} \;=\; \operatorname{codim} \mathfrak{p}A(U) \,.$$

Proof. This result on the codimension ($=$ height) of prime ideals is equivalent to the corresponding statement for polynomial algebras which is easily reduced to the familiar case of a finite number of indeterminates. $\square$

From 1.5 and 1.3 we see that for a noetherian ring A every prime ideal in $A(U)$ has finite codimension. Moreover:

1.6 Proposition *A is noetherian if and only if $A(U)$ is noetherian. If this is the case, $\dim A = \dim A(U)$.*

Proof. Assume that A is noetherian. Then $A(U)$ clearly is noetherian if the family of indeterminates is finite. Now consider the general case and let $\mathfrak{P}$ be any prime ideal of $A(U)$. $\mathfrak{P}$ is contained in a maximal ideal,

---

[1] We emphasize that $\mathfrak{P}' \cap A[U]$ denotes the pre-image of $\mathfrak{P}'$ with respect to the (not necessarily injective) homomorphism $A[U] \to A'[U]$. Similar conventions will be used throughout the notes.

which is extended by 1.3 and has finite codimension by 1.5. Thus $\mathfrak{P}$ has finite codimension, too. For every finite subset $J \subseteq I$ we denote by $\mathfrak{P}_J$ the (finitely generated) prime ideal $(\mathfrak{P} \cap A(U_i : i \in J))A(U)$. The $\mathfrak{P}_J$ form an inductively ordered set of prime ideals in $A(U)$ having $\mathfrak{P}$ for their union. Since $\mathfrak{P}$ has finite codimension there is some $\mathfrak{P}_J$ equal to $\mathfrak{P}$. Thus $\mathfrak{P}$ is finitely generated. By Cohen's Theorem $A(U)$ is noetherian. The converse is obvious. The statement about the dimensions follows by 1.5 and 1.3. $\square$

For instance, 1.6 implies that A is artinian if and only if $A(U)$ is artinian. For any A-module V let

$$V(U) \;=\; V(U_i : i \in I) \;:=\; A(U) \otimes_A V \;=\; V[U]_S \,.$$

The main properties of modules related to Kronecker extensions will be studied in the next section.

Supplements

1. Let V be an A-module. The c o n t e n t $C(G)$ of a polynomial $G \in V[U] = A[U] \otimes_A V$ with coefficients in V is defined to be the submodule of V generated by the coefficients of G. For a primitive polynomial $F \in A[U]$ the following holds:
$$C(FG) \;=\; C(G) \,.$$
(The problem is easily reduced to the case of one variable by substituting the U_i by suitable powers of one variable. Moreover one may assume that A is a field.) Applications:
(1)   Primitive polynomials are non-zero-divisors on $V[U]$. This means that the canonical homomorphism $V[U] \to V(U)$ is an injection.
(2)   Let $F_j \in A[U]$, $j \in J$, be polynomials which are linearly independent modulo $\mathfrak{m}$ for every $\mathfrak{m} \in \mathrm{Spm}\,A$. If $\Sigma_{j \in J} F_j v_j = 0$ in $V(U)$ for elements $v_j \in V$ then $v_j = 0$ for all $j \in J$. More generally: If W is a submodule of V and $\Sigma_{j \in J} F_j v_j \in W(U)$ then $v_j \in W$ for all $j \in J$.

2. Let V_j, $j \in J$, be an arbitrary family of submodules of an A-module V. Then
$$\bigcap_{j \in J} V_j(U) \;=\; A(U)(\bigcap_{j \in J} V_j) \;\subseteq\; V(U) \,.$$
(For an arbitrary family W_j, $j \in J$, of A-modules the canonical homomorphism
$$(\textstyle\prod_{j \in J} W_j)(U) \;\longrightarrow\; \prod_{j \in J} W_j(U)$$
is injective.) Applications:
(1)   If $\mathfrak{a} \subseteq A$ is a radical ideal, so is $\mathfrak{a}A(U) \subseteq A(U)$. In particular, if A is reduced, so is $A(U)$.
(2)   The extension $\mathfrak{m}_A A(U)$ of the Jacobson radical $\mathfrak{m}_A$ of A is the Jacobson radical $\mathfrak{m}_{A(U)}$ of $A(U)$.

3. Let V be an A-module. Then
$$\mathrm{Ass}_{A[U]} V[U] = \{\mathfrak{p}A[U] : \mathfrak{p} \in \mathrm{Ass}_A V\}, \; \mathrm{Ass}_{A(U)} V(U) = \{\mathfrak{p}A(U) : \mathfrak{p} \in \mathrm{Ass}_A V\} \,.$$

4. Let U_i, $i \in I$, and V_j, $j \in J$, be disjoint families of indeterminates over A. Then there is a canonical isomorphism
$$A(U)(V) \;\longleftrightarrow\; A(U;V) \,.$$

(It's easy to write down canonical homomorphisms inverse to each other.)

5. A is normal (that is: an integral domain being integrally closed in its fields of fractions) if and only if $A(U)$ is normal.

More generally: Let A be an integral domain and A' the integral closure of A in its field of fractions K. Then $A'(U)$ is the integral closure of $A(U)$ in its field of fractions $(= K(U))$.

6. Assume $I \neq \emptyset$. For elements $a, b \in A$ the ideals aA and bA coincide if and only if a and b are associated in $A(U)$. (If $a = rb$ and $b = sa$, then $(U+r)b = (sU+1)a$.)

7. Let A be an integral domain and $a, b \in A$. When considering quadratic forms and their discriminants the following is useful: a and b differ (multiplicatively) by the square of a unit in A if and only if the same is true in $A(U)$. (Assume $a \neq 0, b \neq 0$. Let F and G be primitive polynomials in $A[U]$ such that $aG^2 = bF^2$. Then $aA = bA$, i.e. there exists a unit e in A with $a = eb$. We have to prove that e actually is a square in A. For that it suffices to show that F and G generate the same ideal in $A[U]$. This can be checked in $A(V)[U]$ where V_i, $i \in I$, is a new set of indeterminates corresponding to the U_i, $i \in I$. But from $eG^2 = F^2$ and $eG(V)^2 = F(V)^2$ we get $(eG(V)G - F(V)F)(eG(V)G + F(V)F) = 0$.)

8.† Let $A \to A'$ be a homomorphism of commutative rings, U_i, $i \in I$, a non-empty family of indeterminates over A and $h : A' \otimes_A A(U) \to A'(U)$ the canonical homomorphism. This supplement provides some additional remarks on the concordance rule 1.4.

a) Let $A' := A[X_j]_{j \in J}$, $J \neq \emptyset$. Then h is not bijective.

b) Let $A \to A'$ be a field extension. Then h is bijective if and only if $A \to A'$ is algebraic. (For instance, consider a field K between A and A' such that K is purely transcendental over A and A' is algebraic over K.)

c) Let A be a noetherian integral domain and $A' = K$ its field of quotients. Then h is bijective if and only if $\dim A \leq 1$. (Let $\dim A = 1$. One has to show that $K \otimes_A A(U) = K[U]_S (\subseteq K(U))$ is a field. Assume that there is a prime ideal $\mathfrak{M} \neq 0$ in $K[U]_S$. Then $\mathfrak{P} := \mathfrak{M} \cap A[U]_S$ is a prime ideal $\neq 0$ in $A[U]_S = A(U)$ and therefore by 1.6 and 1.3 of the form $\mathfrak{m}A(U)$ with $\mathfrak{m} \in \mathrm{Spm}\, A$. Contradiction.

Let $\dim A \geq 2$. Instead of A we consider the integral closure of A in K. Then we may assume that A is a Krull domain of dimension ≥ 2, which contains a regular sequence a, b with $Aa + Ab \neq A$. Then the polynomial $F := a + bU_{i_0}$ (for an arbitrary $i_0 \in I$) is prime in $A[U]$ but not primitive. F is not invertible in $K[U]_S$.)

d) Let A be noetherian and $\dim A \leq 1$. If for every prime ideal $\mathfrak{p}' \in \mathrm{Spec}\, A'$ the extension $Q(A/A \cap \mathfrak{p}') \subseteq Q(A'/\mathfrak{p}')$ of quotient fields is algebraic, then h is bijective. In particular, h is bijective if A is noetherian of dimension ≤ 1 and if $A' = A_N$ where N is any multiplicative system in A. (Consider a prime ideal $\mathfrak{P}' \subseteq A'[U]$ disjoint to S. One has to show that $\mathfrak{P}'A'(U) \neq A'(U)$. Let $\mathfrak{p}' := A' \cap \mathfrak{P}'$ and $\mathfrak{p} := A \cap \mathfrak{p}'$. Switching to residues modulo $\mathfrak{p}$ and modulo $\mathfrak{p}'$ one reduces to the case $\mathfrak{p} = 0$ and $\mathfrak{p}' = 0$, i.e. $A \subseteq A'$ is an extension of integral domains. Switching to the quotient fields K and $K' = K \otimes_A A'$ of A and A' one reduces to a case covered by c).)

(The fact that Kronecker extensions are not universally compatible with base ring extensions means that they cannot be applied directly in a geometric context.)

2 Modules and Kronecker Extensions

As in the previous section A always denotes a commutative ring. We denote by U_i, $i \in I$, a given system of indeterminates, by $A(U)$ the corresponding Kronecker extension and by $V(U)$ the extended module $A(U) \otimes_A V$ of an A-module V.

2.1 Theorem *Let V be a locally cyclic A-module and let $v_1, \ldots, v_m$ be elements of V. Furthermore, let $F_1, \ldots, F_m$ be polynomials in $A[U]$ the residue classes of which in $(A/\mathfrak{m})[U]$ are linearly independent over $A/\mathfrak{m}$ for every $\mathfrak{m} \in \operatorname{Spm} A$. The following assertions are equivalent:*
(1) $v_1, \ldots, v_m$ *generate V over A.*
(2) $F_1 v_1 + \cdots + F_m v_m$ *generates $V(U)$ over $A(U)$.*
In particular, $V(U)$ is a cyclic $A(U)$-module if V is finite over A.

Proof. (1) follows from (2) using the fact that $A \to A(U)$ is faithfully flat.

(1) implies (2). It is enough to prove that $x := F_1 v_1 + \cdots + F_m v_m$ generates $V(U)$ modulo $\mathfrak{m}V(U)$ for any maximal ideal $\mathfrak{m}$ of A. Thus we may substitute A by any of its residue fields k. By assumption on V, $V(U)$ is then of dimension ≤ 1 over $k(U)$ and hence generated by one of the elements $v_1, \ldots, v_m$, say by v_1. There are elements $a_2, \ldots, a_m$ of k such that $v_i = a_i v_1$, $i = 2, \ldots, m$. Then $x = F v_1$, $F := F_1 + a_2 F_2 + \cdots + a_m F_m$. By assumption on $F_1, \ldots, F_m$ we have $F \neq 0$ and hence $V(U) = k(U)x$. □

The assumption on $F_1, \ldots, F_m$ in 2.1 is equivalent to the following one: $F_1, \ldots, F_m$ generate a free direct A-summand of $A[U]$ of rank m.

For applications one often chooses simply indeterminates $U_1, \ldots, U_m$ for $F_1, \ldots, F_m$. Then

$$U_1 v_1 + \cdots + U_m v_m$$

will be a generating element of $V(U)$, a so-called g e n e r a l g e n e r a t o r of $V(U)$ in the line of Kronecker's method of indeterminates. Another way could be to use the element

$$U^{n_1} v_1 + \cdots + U^{n_m} v_m$$

with just one indeterminate U and pairwise different exponents $n_1, \ldots, n_m$.

2.2 Corollary *Assume $I \neq \emptyset$. Equivalent are:*
(1) V *is a finite projective A-module of rank 1.*
(2) $V(U)$ *is a free $A(U)$-module of rank 1.*

Proof. (2) implies (1) for any faithfully flat change of rings. (1) implies (2) simply by Theorem 2.1. □

It is worth noticing that the technique in Theorem 2.1 can be generalized in natural ways. To formulate one result in that direction we introduce the following notation: Let V be a finite A-module; then

$$\lambda = \lambda(V) = \lambda_A(V) := \min_{\mathfrak{m} \in \mathrm{Spm}\, A} (\dim_{A/\mathfrak{m}} V/\mathfrak{m}V),$$
$$\mu = \mu(V) = \mu_A(V) := \max_{\mathfrak{m} \in \mathrm{Spm}\, A} (\dim_{A/\mathfrak{m}} V/\mathfrak{m}V).$$

2.3 Theorem *Let V be an A-module generated by the elements $v_1, \ldots, v_m$. Furthermore, let (F_{ij}) be a $(m \times \mu)$-matrix with entries from $A[U]$. Assume that for every s with $\lambda \leq s \leq \mu$ there are s columns in (F_{ij}) all the s-minors of which are linearly independent modulo $\mathfrak{m}$ for every $\mathfrak{m} \in \mathrm{Spm}\, A$. Then the μ elements*

$$F_{1j}v_1 + \cdots + F_{mj}v_m, \quad j = 1, \ldots, \mu,$$

generate the $A(U)$-module $V(U)$.

Proof. As in the proof of Theorem 2.1 one is easily led to the case that A is a field. Then the rest of the proof is an exercise in linear algebra. □

The easiest way to construct a matrix (F_{ij}) for 2.3 is to have different indeterminates U_{ij} for F_{ij}. On the other hand, one can always construct matrices of the type wanted by choosing for F_{ij} suitable monomials $U^{n_{ij}}$ in just one indeterminate U over A.

2.4 Corollary *Assume $I \neq \emptyset$. Equivalent are:*
(1)   *V is a finite projective A-module of rank r.*
(2)   *$V(U)$ is a free $A(U)$-module of rank r.*

Theorem 2.3 can be partially generalized in the following way.

2.5 Theorem *Assume $I \neq \emptyset$. For any finitely representable $A(U)$-module W the number $\mu_{A(U)}(W)$ is the minimal number of generators of W.*

Proof. By induction on $\mu = \mu_{A(U)}(W)$ it is sufficient to find an element $w \in W$ such that $\mu_{A(U)}(W/A(U)w) = \mu - 1$ (if $\mu > 0$). Let $W = A(U)w_1 + \cdots + A(U)w_m$. We are going to construct a linear combination $w = F_1 w_1 + \cdots + F_m w_m$, $F_i \in A[U]$, such that for every $\mathfrak{m} \in \mathrm{Spm}\, A$ with $\dim W/\mathfrak{m}W = \mu$ the residue class $\bar{w} \in W/\mathfrak{m}W$ is $\neq 0$.

There is, as we shall see, an integer s with the following property: For any $\mathfrak{m} \in \mathrm{Spm}\, A$ with $\dim W/\mathfrak{m}W = \mu$ there is a $J \subseteq \{1, \ldots, m\}$ with $|J| = \mu$ and with polynomials G_{ij}, $G \in (A/\mathfrak{m})[U]$, $\deg G_{ij} \leq s$, $G \neq 0$ and

$$\bar{w}_j = \sum_{i \in J} \frac{G_{ij}}{G} \bar{w}_i, \quad j = 1, \ldots, m,$$

in $W/\mathfrak{m}W$. Then let $F \in A[U]$ be a homogeneous primitive polynomial of degree $> s$ and set $F_j := F^{j-1}$, $j = 1, \ldots, m$. In $W/\mathfrak{m}W$ the coefficient of $\bar{w} = \bar{F}_1 \bar{w}_1 + \cdots + \bar{F}_m \bar{w}_m$ at $\bar{w}_i$ is $G^{-1}\Sigma_{j=1}^m \bar{F}_j G_{ij} \neq 0$ where $i \in J$ is chosen in such a way that $(G_{i1}, \ldots, G_{im}) \neq 0$ (which is possible because $(G_{ij}) \neq 0$).

It remains to show how to find s, J, G_{ij}, G. One way to do it is to represent W as a residue class module of a free module $A(U)^m$ with basis $e_1, \ldots, e_m$ modulo a submodule generated by (finitely many) elements $y_k = \Sigma_{j=1}^m H_{jk} e_j$, $k = 1, \ldots, r$, $H_{jk} \in A[U]$. Consider any $\mathfrak{m} \in \operatorname{Spm} A$ with $\dim W/\mathfrak{m}W = \mu$. There is a subset J of $\{1, \ldots, m\}$ such that $\bar{w}_i$, $i \in J$, is a basis of $W/\mathfrak{m}W$ and a subset K of $\{1, \ldots, r\}$ such that $\bar{e}_i$, $i \in J$, and $\bar{y}_k$, $k \in K$, together form a basis of $(A/\mathfrak{m})(U)^m$. Then choose for G resp. G_{ij} the residue classes of the determinant resp. of suitable minors of the matrix built with e_i, $i \in J$, and y_k, $k \in K$, as columns. There are only finitely many choices.   □

2.6 Corollary *Assume $I \neq \emptyset$. Any finite projective module over $A(U)$ with rank is free. In particular, (finite) stably free modules over $A(U)$ are free. The group $\operatorname{Pic} A(U)$ of classes of finite projective $A(U)$-modules of rank one is trivial.*

Finally we broach the problem of determining algebra generators of finite commutative algebras when Kronecker extensions are applied.

Consider a finite commutative algebra B over a field k. Let K be any field extension of k with infinitely many elements. Then the minimal number of K-algebra generators of $B_{(K)} = K \otimes_k B$ is independent of the choice of K; we denote it by

$$\nu = \nu_k(B)$$

and call it the t r u e n u m b e r of k-algebra generators of B. (That ν is independent of the choice of the infinite field K follows from the fact that the number of algebra generators does not decrease when field extensions of K are applied.)

For an arbitrary finite commutative A-algebra B we define

$$\nu = \nu_A(B) := \max_{\mathfrak{m} \in \operatorname{Spm} A} (\nu_{A/\mathfrak{m}}(B/\mathfrak{m}B)).$$

2.7 Theorem *Let $B = A[x_1, \ldots, x_m]$ be a finite commutative A-algebra and let (F_{ij}) be a $(m \times \nu)$-matrix with entries from $A[U]$, $\nu = \nu_A(B)$. Assume that for every $s = 1, \ldots, \nu$ there are s columns in (F_{ij}) all the s-minors of which are linearly independent modulo $\mathfrak{m}$ for every $\mathfrak{m} \in \operatorname{Spm} A$. Then the ν elements*

$$y_j := F_{1j}x_1 + \cdots + F_{mj}x_m, \quad j = 1, \ldots, \nu,$$

generate the $A(U)$-algebra $B(U)$.

Proof. Again we may assume that $A = k$ is a field and even that k is algebraically closed. (The concordance rule is being used.) In addition, we may switch to the local components of B, i.e. we may assume that B is local with maximal ideal $\mathfrak{M}$. We replace x_i by $x_i - a_i$, where $a_i \in k$ is the residue class of x_i, and thus may finally assume that the x_i generate $\mathfrak{M}$ as a B-module. But then $y_1, \ldots, y_\nu$ generate $\mathfrak{M}(U)$ over $B(U)$ by 2.3, and $B(U) = k(U)[y_1, \ldots, y_\nu]$. $\qquad\qquad\qquad\qquad\qquad\qquad\qquad\qquad$ $\square$

In particular the situation of 2.7 can always be realized using a Kronecker extension by one indeterminate.

2.8 Corollary *Let B be a finite commutative A-algebra. If $I \neq \emptyset$ then the $A(U)$-algebra $B(U)$ can be generated by $\nu_A(B)$ elements.*

For a generalization see Supplement 9.

Supplements

1. Let $F_1, \ldots, F_r \in A[U]$ be polynomials such that $C(F_1) + \cdots + C(F_r)$ is a projective A-ideal. Then
$$(F_1, \ldots, F_r)A(U) = (C(F_1) + \cdots + C(F_r))A(U).$$
Moreover, if $I \neq \emptyset$, this ideal is principal (generated by $F = G_1F_1 + \cdots + G_rF_r$ where $G_1, \ldots, G_r$ are suitably chosen monomials; use Theorem 2.1).

2. (*Prüfer* and *Bezout domains*) Let A be an integral domain. A is called a P r ü f e r (resp. B e z o u t) d o m a i n if finitely generated ideals in A are projective (resp. free) A-modules. If $I \neq \emptyset$ then the following assertions are equivalent:
(1) A is a Prüfer domain.
(2) $A(U)$ is a Bezout domain.
(3) $A(U)$ is a Prüfer domain.
If these equivalent conditions are fulfilled then every finitely generated ideal of $A(U)$ is extended from A.

3. (*Dedekind domains*) Assume that $I \neq \emptyset$. Then the following assertions are equivalent:
(1) A is a Dedekind domain.
(2) $A(U)$ is a principal ideal domain.
(3) $A(U)$ is a Dedekind domain.
If these equivalent conditions are fulfilled,
$$\mathfrak{a} \mapsto \mathfrak{a}A(U)$$
is an isomorphism from the lattice of ideals of A to the lattice of ideals of $A(U)$.

4. (*Frobenius algebras*) Let B be a finite projective commutative A-algebra. B is called a q u a s i - F r o b e n i u s (resp. F r o b e n i u s) a l g e b r a over A if the r e l a t i v e   d u a l i z i n g   m o d u l e
$$E := \operatorname{Hom}_A(B, A)$$
is a projective (resp. free) B-module (necessarily of rank 1). If $I \neq \emptyset$ then the following assertions are equivalent: (1) B is a quasi-Frobenius algebra over A. (2) $B(U)$ is a Frobenius algebra over $A(U)$. (3) $B(U)$ is a quasi-Frobenius algebra over $A(U)$. $(\operatorname{Hom}_{A(U)}(B(U), A(U)) = E(U).)$

5.† Let A be a Krull domain, $I \neq \emptyset$. Then $A(U)$ is a Krull domain and the canonical map $\mathrm{Cl}\,(A) \to \mathrm{Cl}\,(A(U))$ of divisor class groups is surjective with kernel $\mathrm{Pic}\,A$, cf. [4], Ch. VII, § 1, no. 10, i. e. there is a canonical exact sequence

$$1 \longrightarrow \mathrm{Pic}\,A \longrightarrow \mathrm{Cl}\,(A) \longrightarrow \mathrm{Cl}\,(A(U)) \longrightarrow 0\,.$$

In particular, $\mathrm{Pic}\,A(U)$ is trivial. $A(U)$ is factorial if and only if A is locally factorial. ($\mathrm{Pic}\,A(U) = 1$ follows also via 2.6.)

6.† The construction of Kronecker extensions can be generalized in an obvious way. As before let U_i, $i \in I$, be a system of indeterminates over A and assume $I \neq \emptyset$. Let Y be an arbitrary subset of $\mathrm{Spec}\,A$. Then the set $S = S_Y$ of polynomials $F \in A[U]$ with $\mathrm{C}(F) \not\subseteq \mathfrak{p}$ for every $\mathfrak{p} \in Y$ is a saturated multiplicative system in $A[U]$ (containing the system of primitive polynomials). The ring of fractions $A[U]_S$ will be denoted by $A_Y(U)$. It is flat over A. The space $\mathrm{Spm}\,A_Y(U)$ consists of the prime ideals $\mathfrak{p}A_Y(U)$ where $\mathfrak{p} \in \mathrm{Spec}\,A$ is maximal with respect to the following property: Any finitely generated subideal $\mathfrak{a} \subseteq \mathfrak{p}$ is contained in some prime ideal of Y. If A is noetherian then $\mathrm{Spm}\,A_Y(U)$ consists of the prime ideals $\mathfrak{p}A_Y(U)$ where $\mathfrak{p}$ is maximal in Y. The analogon of Theorem 2.5 holds. For $Y = \mathrm{Spec}\,A$ we get the Kronecker extension $A(U)$ of A.

7.† Let A be a Krull domain which is not a field and assume $I \neq \emptyset$. By Y_1 we denote the prime ideals of codimension 1 in A. The polynomials of S_{Y_1} are called 1 - p r i m i t i v e. Then the ring $A_{Y_1}(U)$, defined in Suppl. 6, is a principal ideal domain, and the canonical homomorphism $\mathrm{Div}\,(A) \to \mathrm{Div}\,(A_{Y_1}(U))$ of groups of divisors is bijective. For a Dedekind domain A we have, of course, $A_{Y_1}(U) = A(U)$, cf. Suppl. 3 above.

(This construction goes back to Kronecker and was described more precisely by J. König in [27], IX, §§7,8, who thereby gave an interpretation of Kummer's *ideal numbers* in terms of polynomials: A divisor $d \in \mathrm{Div}\,(A)$ is represented by a polynomial $F \in A[U]$ which defines the divisor d in $\mathrm{Div}\,(A_{Y_1}(U))$.)

The construction of $A_{Y_1}(U)$ should not be confounded with the construction of $A[U]_P$ where P is the multiplicative system generated by the prime elements of $A[U]$. $A[U]_P$ has the same divisor class group as A and is a Dedekind domain (or a field) if $I \neq \emptyset$. This depends mainly on the following statement the proof of which we indicate for the sake of completeness: *Every prime ideal* $\mathfrak{P} \subseteq A[U], I \neq \emptyset$, *of codimension* > 1 *contains a prime element*. We may assume right away $I = \{1\}$. Let $\mathfrak{p} := \mathfrak{P} \cap A$. Then $\mathrm{codim}\,\mathfrak{p} \geq 1$. If $\mathrm{codim}\,\mathfrak{p} > 1$ and $a, b \in \mathfrak{p}$ are chosen coprime $a + bU \in \mathfrak{P}$ is prime. Now let $\mathrm{codim}\,\mathfrak{p} = 1$. Choose $F = a_0 + a_1U + \cdots + a_nU^n \in \mathfrak{P}$ such that F generates $\mathfrak{P}A_{\mathfrak{p}}[U]/\mathfrak{p}A_{\mathfrak{p}}[U]$, $a_n \notin \mathfrak{p}$. Let $\mathfrak{p}_1, \ldots, \mathfrak{p}_r$ be the prime ideals of codimension 1 in A containing $a_0, a_1, \ldots, a_n$ and $\mathfrak{p}_1, \ldots, \mathfrak{p}_r, \ldots, \mathfrak{p}_s$ those containing $a_1, \ldots, a_n$. If $a \in \mathfrak{p}_{r+1} \cap \cdots \cap \mathfrak{p}_s \cap \mathfrak{p}$, $a \notin \mathfrak{p}_1 \cup \cdots \cup \mathfrak{p}_r$, then $G = a + F \in \mathfrak{P}$ is prime because G is 1-primitive and prime over the quotient field K of A. (Any decomposition of G over K induces a decomposition in $A_{\mathfrak{p}}[U]$ modulo $\mathfrak{p}A_{\mathfrak{p}}[U]$.))

8.† Let A be a Krull domain, K its field of fractions, L a finite extension field of K and B the integral closure of A in L (a Krull domain, too). A polynomial $F \in B[U]$ is 1-primitive if and only if its norm $\mathrm{N}_{A[U]}^{B[U]}(F)$ is 1-primitive.

(For the definition of 1-primitive polynomials see Suppl. 7. The proof is done by localizing A with respect to prime ideals $\mathfrak{p}$ of codimension 1 in A and looking at the Kronecker extensions $A_{\mathfrak{p}}(U) \subseteq B_{\mathfrak{p}}(U)$.)

The criterion given here served Kronecker and König to define 1-primitive polynomials over B in case that the subdomain A is factorial, thus reducing the concept to the one of Gauß for polynomials over factorial domains.)

9. Assume $I \neq \emptyset$ and let C be a finite and finitely represented commutative $A(U)$-algebra. Then $\nu_{A(U)}(C)$ is the minimal number of $A(U)$-algebra generators of C. (The proof runs along the same lines as that of 2.5.)

10.† Let B be an arbitrary finite commutative A-algebra. The invariant $\nu = \nu_A(B)$, which we referred to in Theorem 2.7, can be computed using the module $\Omega_A(B)$ of Kähler differentials: One has $\nu = \max(1, \mu_B(\Omega_A(B)))$ if the structure homomorphism $A \to B$ is not surjective, and $\nu = 0$ otherwise. (For the proof we may by definition of ν assume that $A = k$ is an infinite field. Then the result is Lemma (4.2) in [37]. For the reader's convenience we give an outline of the proof. First we may assume that k is even algebraically closed. Looking at the local components of B and using the Chinese Remainder Theorem we may furthermore assume that B is a local k-algebra. But in this case

$$\mu_B(\Omega_k(B)) = \dim_k(\mathfrak{m}_B/\mathfrak{m}_B^2) = \nu_k(B).)$$

11. Assume $I \neq \emptyset$. Let $B = A[x_1, \ldots, x_m]$ be a finite commutative A-algebra and n a positive integer. For every $\mathfrak{m} \in \operatorname{Spm} A$ assume that the fibre $B/\mathfrak{m}B$ has local components of embedding dimension $\leq n$ and residue fields which are separable over $A/\mathfrak{m}$. Then we have $\nu_A(B) \leq n$, and there are linear combinations $y_1, \ldots, y_n$ of $x_1, \ldots, x_m$ with coefficients in $A[U]$ such that $B(U) = A(U)[y_1, \ldots, y_n]$.

12.† Let B be a finite projective commutative A-algebra. B is called s e p a r a b l e (or é t a l e) over A if B is u n r a m i f i e d over A, i.e. if the module $\Omega_A(B)$ of Kähler differentials is the zero-module (or equivalently, if $E = \operatorname{Hom}_A(B, A)$ is generated as a B-module by the ordinary trace tr_A^B, or again equivalently if the fibres $A/\mathfrak{m} \to B/\mathfrak{m}B$ are separable for every $\mathfrak{m} \in \operatorname{Spm} A$). To have a smooth formulation in the following let B have rank r over A.

If B is separable over A, so is $B(U)$ over $A(U)$. If furthermore $I \neq \emptyset$ then by Suppl. 11 the $A(U)$-algebra $B(U)$ has a primitive element, i.e. there is a monic (separable) polynomial α of degree r in one indeterminate X over $A(U)$ such that $B(U) = A(U)[X]/(\alpha)$.

13.† Let $B = A[x_1, \ldots, x_m]$ be a finite projective commutative A-algebra with rank and $F_1, \ldots, F_m$ polynomials in $A[U]$ which are linearly independent modulo $\mathfrak{m}$ for every $\mathfrak{m} \in \operatorname{Spm} A$. (For example, $F_i := U_i$, or $F_i := U^{n_i}$ with just one indeterminate U and pairwise different exponents $n_1, \ldots, n_m$.) Let
$$x := F_1 x_1 + \cdots + F_m x_m \in B(U)$$
and denote the characteristic polynomial of x by $\chi \in A(U)[X]$. For every $\mathfrak{m} \in \operatorname{Spm} A$ such that the residue fields of $B/\mathfrak{m}B$ are separable over $A/\mathfrak{m}$ let

$$\bar{\chi} = \prod_{k=1}^{r} P_k^{e_k}$$

be the canonical prime factor decomposition of the residue class $\bar{\chi}$ of χ in the polynomial ring $(A/\mathfrak{m})(U)[X]$. Then the monic prime polynomials P_k are the minimal polynomials of x in the residue fields of $(B/\mathfrak{m}B)(U)$ over $(A/\mathfrak{m})(U)$ and the e_k are the multiplicities of the corresponding local components of $(B/\mathfrak{m}B)(U)$, which, of course, have the same multiplicities as the local components of $B/\mathfrak{m}B$, in which one might have been primarily interested. (One can assume $A = A/\mathfrak{m}$

and $B = B/\mathfrak{m}B$ local. In this case $\chi = P^e$, where P is the minimal polynomial of the residue class of x in the residue field of B, which generates this field by Theorem 2.7. In general, the separability condition is necessary: Let K be a field of prime characteristic p such that there is an element $a \in K \smallsetminus K^p$. Consider $B := K[X,Y]/(X^p, Y^p - a)$ over $A := K$.)

3 Numerical Monoids

The discussion of the grading of a polynomial algebra $A[T_0, \ldots, T_n]$ given by positive w e i g h t s $\gamma_0, \ldots, \gamma_n$ of the indeterminates $T_0, \ldots, T_n$ naturally leads to questions about the monoid generated by the weights. Therefore, in this section and the next one we will introduce some concepts and propositions about (additive) numerical monoids. For general as well as for historical references we refer to [5], [13] and [2].

We assume throughout that the given *positive* weights $\gamma_0, \ldots, \gamma_n$ are relatively prime, i. e. $\gcd(\gamma_0, \ldots, \gamma_n) = 1$. Then they generate a numerical submonoid of $\mathbb{N}$ which shall be denoted by

$$\mathrm{Mon}(\gamma) \;=\; \mathrm{Mon}(\gamma_0, \ldots, \gamma_n)\,.$$

Every numerical submonoid of $\mathbb{N}$ can be described in such a way. Note that a submonoid M of $\mathbb{N}$ is called n u m e r i c a l if and only if the set $\mathbb{N}\smallsetminus M$ of g a p s is finite. Then

$$\vartheta_M \;:=\; |\mathbb{N}\smallsetminus M| \;=\; \mathrm{card}\,(\mathbb{N}\smallsetminus M)$$

is called the d e g r e e o f s i n g u l a r i t y (or the g e n u s) of M. (Usually this number is denoted by δ_M which we are to avoid, however, because we reserve δ for the degrees of polynomials.) A numerical submonoid M of $\mathbb{N}$ defines the monoid algebra $K[M]$ over an (arbitrary) field K, which is the graded subalgebra $\sum_{m \in M} K X^m$ of the polynomial algebra $K[\mathbb{N}] = K[X]$ with $\deg X = 1$. The ring $K[X]$ is the normalization of $K[M]$, and $\vartheta_M = \dim_K K[X]/K[M]$ which motivates the expressions "degree of singularity" (and "genus") for ϑ_M. To ring-theoretic concepts concerning the ideals of $K[M]$ correspond concepts about ideals of M.

Let $M \subseteq \mathbb{N}$ be a numerical submonoid of $\mathbb{N}$. A subset $I \subseteq \mathbb{Z}$ is called an M-i d e a l if $I \smallsetminus \mathbb{N}$ is finite and if $M + I \subseteq I$, where $M + I = \{a + x : a \in M, x \in I\}$. The empty ideal will be called the z e r o - i d e a l. An ideal $I \neq \emptyset$ is also called a f r a c t i o n a l ideal of M. The fractional ideals of M correspond bijectively to the homogeneous fractional ideals of $K[M]$ in its graded quotient ring $K[X, X^{-1}]$. To the usual operations $\mathfrak{a} + \mathfrak{b}$, $\mathfrak{a} \cap \mathfrak{b}$, $\mathfrak{a} \cdot \mathfrak{b}$ and $\mathfrak{a} : \mathfrak{b}$ of fractional ideals $\mathfrak{a}, \mathfrak{b}$ over $K[M]$ correspond the operations $I \cup J$, $I \cap J$, $I + J$ and

$$[I - J] \;:=\; \{z \in \mathbb{Z} : z + J \subseteq I\} \quad \text{(assuming } J \neq \emptyset\text{)}$$

of M-ideals I, J. The m a x i m a l i d e a l of M is

$$M_+ \;:=\; M \smallsetminus \{0\}\,.$$

For any M-ideal I the finite set $I \setminus (M_+ + I)$ is the uniquely determined minimal set of generators of I as an M-ideal. The cardinality of the minimal set $E := M_+ \setminus (M_+ + M_+)$ of generators of the maximal ideal is sometimes called the **embedding dimension** of M. The set E is also the uniquely determined minimal set of generators of the numerical monoid M itself. *Proof*: Of course, any subset of M which generates M contains E. To prove $M = \mathrm{Mon}(E)$, assume the contrary and consider the minimal element $c \in M \setminus \mathrm{Mon}(E)$. Then $c \notin E$, hence $c = a+b$ with $a, b \in M_+$. Being smaller than c, the elements a,b must be in $\mathrm{Mon}(E)$. Hence $c \in \mathrm{Mon}(E)$, too, a contradiction.

The ordinary dual of an ideal $I \neq \emptyset$ is $[M - I]$. As the **adjoint ideal** (or simply **adjoint**) of an M-ideal $I \neq \emptyset$ we define the M-ideal

$$I^\wedge := \{z \in \mathbb{Z} : -z \notin I\}.$$

Obviously, $I^{\wedge\wedge} = I$. The ideal

$$\Omega_M := M^\wedge = \{z \in \mathbb{Z} : -z \notin M\}$$

is called the **dualizing ideal** (or **canonical ideal**) of M. (The corresponding ideal over $K[M]$ is the dualizing module (or canonical module) $\omega_{K[M]}$ of the graded algebra $K[M]$; cf. [6], Section 3.6.) Then

$$I^\wedge = [\Omega_M - I]$$

for all M-ideals $I \neq \emptyset$.

For any ideal $I \neq \emptyset$ let

$$v_I := \min I$$

be called the **order** of I. Then

$$g_I := -v_{I^\wedge}$$

is the so-called **Frobenius number** of I which is nothing else but the maximum in the set of integers not belonging to I. For $M = \mathrm{Mon}(\gamma)$ we will also write

$$g_\gamma := g_M \quad \text{and} \quad \vartheta_\gamma := \vartheta_M$$

for the Frobenius number and the degree of singularity of M. By

$$\vartheta_I := \mathrm{card}\,((\mathbb{N} + v_I) \setminus I) = \mathrm{card}\,\{z \in \mathbb{Z} : z > v_I, z \notin I\}$$

a singularity degree can be defined for I, too. In case $\vartheta_I > 0$, g_I is just the maximum of $(\mathbb{N} + v_I) \setminus I$.

The set

$$T(I) := I^\wedge \setminus (M_+ + I^\wedge),$$

i. e. the minimal set of generators of the adjoint of the ideal $I \neq \emptyset$, will be called the **t y p e   s e t** of I. (Some authors take the set $-T(I)$ for the type set of I.) Then $-g_I$ is the smallest element in $T(I)$. The cardinality of $T(I)$ is the so-called **t y p e** of I, denoted by t_I . If $t_M = 1$, i. e. if the dualizing ideal Ω_M is a principal ideal, then M is called **s y m m e t r i c**. (This means that $K[M]$ is a Gorenstein ring. See also Supplement 1.)

The characteristic function of an M-ideal I is the Hilbert function of the corresponding homogeneous ideal $\sum_{p \in I} K X^p$. Its generating function

$$\mathcal{P}_I := \sum_{p \in I} Z^p$$

is called the **P o i n c a r é   s e r i e s** (or **H i l b e r t   s e r i e s**) of I. This series is always a rational function of type

$$\mathcal{P}_I = \frac{\mathcal{G}_I}{1 - Z}$$

with a Laurent polynomial $\mathcal{G}_I \in \mathbb{Z}[Z, Z^{-1}]$. If $I \neq \emptyset$, then $\mathcal{G}_I$ is the polynomial

$$\mathcal{G}_I = \left(\sum_{p \in I, v_I \leq p < g_I} Z^p \right)(1 - Z) + Z^{g_I + 1}$$

with the lower degree v_I and the upper degree $g_I + 1$. Note that

$$\mathcal{G}_I(1) = 1.$$

Another consequence is the equation

$$\vartheta_I = \mathcal{G}_I'(1) - v_I,$$

where $\mathcal{G}_I'$ is the derivative of the Laurent polynomial $\mathcal{G}_I$, in particular $\vartheta_M = \mathcal{G}_M'(1) = \mathcal{G}_M'(1)/\mathcal{G}_M(1)$. Furthermore,

$$\mathcal{P}_{I^*} = -\mathcal{P}_I(\tfrac{1}{Z}), \quad \mathcal{G}_{I^*} = Z\mathcal{G}_I(\tfrac{1}{Z}).$$

Another way of presenting $\mathcal{P}_I$ is to choose an element $m \in M_+$ and to compute the numbers

$$u_i = q_i m + i := \min(I \cap (\mathbb{Z}m + i))$$

for $i = 0, \ldots, m - 1$. Then

$$\mathcal{P}_I = \frac{1}{1 - Z^m} \sum_{i=0}^{m-1} Z^{u_i}$$

because of $I = \biguplus_{i=0}^{m-1}(I \cap (\mathbb{Z}m + i))$, which implies

$$v_I = \min_{0 \leq i \leq m-1} u_i, \quad g_I = \max_{0 \leq i \leq m-1} u_i - m,$$

$$\vartheta_I + v_I = \sum_{i=0}^{m-1} q_i = \tfrac{1}{m}\left(\sum_{i=0}^{m-1} u_i \right) - \tfrac{m-1}{2}.$$

These formulas are often used to compute Frobenius numbers and degrees of singularities, see Supplement 6 for an example. The family u_i, $i = 0, \ldots, m - 1$, is called the Apéry basis or standard basis of I with respect to m.

The Frobenius number and the degree of singularity of the monoid M do not depend on a special representation $M = \mathrm{Mon}(\gamma)$, $\gamma = (\gamma_0, \ldots, \gamma_n)$. The opposite is the case for the concepts we turn to, now. One should keep in mind, though, that the set $\{\gamma_0, \ldots, \gamma_n\}$ always contains the uniquely determined minimal set of generators of M.

For $\alpha \in \mathbb{Z}^{n+1}$ let

$$\langle \alpha, \gamma \rangle := \alpha_0 \gamma_0 + \cdots + \alpha_n \gamma_n.$$

The sum $\gamma_0 + \cdots + \gamma_n = \langle (1, \ldots, 1), \gamma \rangle$ will be used in several formulas. For convenience we abbreviate it by

$$|\gamma| := \gamma_0 + \cdots + \gamma_n.$$

For $m \in \mathbb{Z}$,

$$\mathrm{N}_m(\gamma) := \{\nu \in \mathbb{N}^{n+1} : \langle \nu, \gamma \rangle = m\}$$

denotes the set of all $(n + 1)$-tuples of non-negative integers representing m. Then

$$\Gamma(m) = \Gamma(m; \gamma) := \sum_{\nu, \mu \in \mathrm{N}_m(\gamma)} \mathbb{Z}(\nu - \mu) \subseteq \mathbb{Z}^{n+1}$$

is a subgroup of the group

$$\Gamma = \mathrm{Syz}(\gamma) := \{\alpha \in \mathbb{Z}^{n+1} : \langle \alpha, \gamma \rangle = 0\}$$

of all syzygies of $\gamma_0, \ldots, \gamma_n$. In case $[\Gamma : \Gamma(m)] < \infty$ we call m a sating number (with respect to γ). The set

$$\mathrm{Sat}(\gamma) := \{m \in M : m \text{ sating for } \gamma\}$$

is an $\mathrm{Mon}(\gamma)$-ideal since $\Gamma(m) \subseteq \Gamma(x + m)$ for $x \in \mathrm{Mon}(\gamma)$, $m \in \mathbb{N}$. The ideal $\mathrm{Sat}(\gamma)$ is not empty because $\Gamma(m) = \Gamma$ for $m := \mathrm{lcm}(\gamma_0, \ldots, \gamma_n)$. The Frobenius number $g_{\mathrm{Sat}(\gamma)}$ of $\mathrm{Sat}(\gamma)$ will be abbreviated by

$$\mathrm{sat}_\gamma.$$

In case $\Gamma(m) = \Gamma$, we call m a very sating number (with respect to γ). The set

$$\mathrm{VSat}(\gamma) := \{m \in M : m \text{ very sating for } \gamma\}$$

is also a non-empty $\mathrm{Mon}(\gamma)$-ideal. Its Frobenius number $g_{\mathrm{VSat}(\gamma)}$ will be denoted by

$$\mathrm{vsat}_\gamma.$$

Always $\text{VSat}(\gamma) \subseteq \text{Sat}(\gamma)$ and consequently $\text{sat}_\gamma \leq \text{vsat}_\gamma$. This inequality may be strict, e.g. for $\gamma = (6,8,9,13)$ one has $\text{sat}_\gamma = 31$ and $\text{vsat}_\gamma = 32$. Another example is provided by $\gamma = (6,7,9,17)$. For any $\gamma = (\gamma_0, \ldots, \gamma_n)$ with $n \leq 2$, the equality $\text{sat}_\gamma = \text{vsat}_\gamma$ holds, see Supplement 14.

More generally, for arbitrary $r \in \mathbb{N}$ the M-ideals

$$\text{Sat}^r(\gamma) := \{m \in M \ : \ \text{rank}(\Gamma/\Gamma(m;\gamma)) \leq r\}$$

can be defined. The Frobenius numbers of these ideals will be denoted by sat_γ^r. One has $\text{Sat}^0(\gamma) = \text{Sat}(\gamma)$ and $\text{Sat}^r(\gamma) = M$ for $r \geq n$.

3.1 Proposition *Let $n \geq 1$. For $\gamma = (\gamma_0, \ldots, \gamma_n)$ and $M := \text{Mon}(\gamma)$ the following holds:*

(1) $\text{VSat}(\gamma) \subseteq \text{Sat}(\gamma) \subseteq M + \gamma_i \subseteq M_+$ *for $i = 0, \ldots, n$.*

(2) $g_\gamma + \max_i \gamma_i \leq \text{sat}_\gamma \leq \text{vsat}_\gamma \leq g_\gamma + \max_{i<j} \text{lcm}(\gamma_i, \gamma_j)$.

Proof. To prove (1) consider $m \in \text{Sat}(\gamma)$. There is an $\alpha = (\alpha_0, \ldots, \alpha_n)$ in $\mathbb{N}_m(\gamma)$ such that $m = \langle \alpha, \gamma \rangle$ and $\alpha_i > 0$; otherwise $\Gamma(m)$ would contain only syzygies $(a_0, \ldots, a_n)$ with $a_i = 0$, which contradicts the condition $[\Gamma : \Gamma(m)] < \infty$. Thus $m - \gamma_i \in M$.

The first two inequalities in (2) follow from (1). To prove the last inequality, observe first that Γ is generated by the regular syzygies $\gamma_j e_i - \gamma_i e_j$, $i \neq j$, where $e_0, \ldots, e_n$ is the standard basis of $\mathbb{Z}^{n+1}$. Let h denote the maximum of the $h_{ij} := \text{lcm}(\gamma_i, \gamma_j)$, $0 \leq i < j \leq n$, and consider any $m > g_\gamma + h$. For a special pair $i < j$ there is an element $\alpha \in \mathbb{N}^{n+1}$ with $m - h_{ij} = \langle \alpha, \gamma \rangle$. But then

$$m = \langle \alpha, \gamma \rangle + h_{ij} = \langle \alpha + \frac{\gamma_j}{\gcd(\gamma_i, \gamma_j)} e_i, \gamma \rangle = \langle \alpha + \frac{\gamma_i}{\gcd(\gamma_i, \gamma_j)} e_j, \gamma \rangle$$

and the difference of these two representations of the number m yields even $\gcd(\gamma_i, \gamma_j)^{-1}(\gamma_j e_i - \gamma_i e_j) \in \Gamma(m)$. Therefore $\Gamma(m)$ contains the regular syzygies. $\square$

The following proposition will be helpful in the handling of sat_γ and vsat_γ.

3.2 Proposition *For $\gamma = (\gamma_0, \ldots, \gamma_n)$ and $\bar{\gamma} := (\gamma_0, \ldots, \gamma_n, \gamma_{n+1})$ with $\gamma_{n+1} \in M := \text{Mon}(\gamma)$ the following holds:*

(1) $\text{Sat}(\bar{\gamma}) = \text{Sat}(\gamma) \cap (M + \gamma_{n+1})$, $\text{sat}_{\bar{\gamma}} = \max(\text{sat}_\gamma, g_\gamma + \gamma_{n+1})$.

(2) $\text{VSat}(\bar{\gamma}) = \text{VSat}(\gamma) \cap (M + \gamma_{n+1})$, $\text{vsat}_{\bar{\gamma}} = \max(\text{vsat}_\gamma, g_\gamma + \gamma_{n+1})$.

In particular, $\text{Sat}(\bar{\gamma}) = \text{Sat}(\gamma)$ if and only if $\text{Sat}(\gamma) \subseteq M + \gamma_{n+1}$, and $\text{VSat}(\bar{\gamma}) = \text{VSat}(\gamma)$ if and only if $\text{VSat}(\gamma) \subseteq M + \gamma_{n+1}$.

Proof. The canonical inclusions $\text{Syz}(\gamma) \subseteq \text{Syz}(\bar{\gamma})$ and $\Gamma(m;\gamma) \subseteq \Gamma(m;\bar{\gamma})$ induce a homomorphism

$$h \ : \ \text{Syz}(\gamma)/\Gamma(m;\gamma) \longrightarrow \text{Syz}(\bar{\gamma})/\Gamma(m;\bar{\gamma}).$$

We fix a representation $\gamma_{n+1} = \varepsilon_0\gamma_0 + \cdots + \varepsilon_n\gamma_n$ with $\varepsilon_i \in \mathbb{N}$. Then $\text{Syz}(\bar{\gamma}) = \text{Syz}(\gamma) \oplus \mathbb{Z}s$ with $s := (\varepsilon_0, \ldots, \varepsilon_n, -1)$. The homomorphism $\mathbb{Z}^{n+1} \to \mathbb{Z}^n$ with $(a_0, \ldots, a_n, a_{n+1}) \mapsto (a_0 + \varepsilon_0 a_{n+1}, \ldots, a_n + \varepsilon_n a_{n+1})$ maps $\text{Syz}(\bar{\gamma})$ into $\text{Syz}(\gamma)$ and $\text{N}_m(\bar{\gamma})$ into $\text{N}_m(\gamma)$, therefore $\Gamma(m; \bar{\gamma})$ into $\Gamma(m; \gamma)$, and hence induces a homomorphism

$$g \ : \ \text{Syz}(\bar{\gamma})/\Gamma(m; \bar{\gamma}) \longrightarrow \text{Syz}(\gamma)/\Gamma(m; \gamma).$$

Because of $g \circ h = \text{id}$ the homomorphism h is injective. This proves $\text{Sat}(\bar{\gamma}) \subseteq \text{Sat}(\gamma)$ and $\text{VSat}(\bar{\gamma}) \subseteq \text{VSat}(\gamma)$. With 3.1(1) we get the inclusions $\text{Sat}(\bar{\gamma}) \subseteq \text{Sat}(\gamma) \cap (M + \gamma_{n+1})$ and $\text{VSat}(\bar{\gamma}) \subseteq \text{VSat}(\gamma) \cap (M + \gamma_{n+1})$. To prove the opposite inclusions it suffices to show that for $m \in M + \gamma_{n+1}$ the homomorphism h is surjective. But an element $m = \alpha_0\gamma_0 + \cdots + \alpha_n\gamma_n + \gamma_{n+1}$ has also the representation $m = (\alpha_0 + \varepsilon_0)\gamma_0 + \cdots + (\alpha_n + \varepsilon_n)\gamma_n$. Therefore $s \in \Gamma(m; \bar{\gamma})$ which means that h is surjective.

The additional formulas about Frobenius numbers follow from the general formula

$$g_{I \cap J} \ = \ \max(g_I, g_J)$$

for arbitrary fractional M-ideals I, J.   □

In some cases monoids can be reduced to simpler ones in a natural way. The most important one of these techniques is the so-called r e d u c t i o n i n c o d i m e n s i o n 1 , originated by S.M. Johnson in [20], which we are going to describe now. Another one will be studied in Section 5.

Again, let $\gamma = (\gamma_0, \ldots, \gamma_n)$, $n \geq 1$, generate the numerical monoid $M = \text{Mon}(\gamma)$. Let us assume that $d \in \mathbb{N}^*$ is a common divisor of $\gamma_1, \ldots, \gamma_n$. Then $\gcd(\gamma_0, d) = 1$. Let

$$\gamma_0' := \gamma_0, \quad \gamma_i' := \gamma_i/d \ \text{for} \ i = 1, \ldots, n, \quad \gamma' := (\gamma_0', \ldots, \gamma_n')$$

and $M' := \text{Mon}(\gamma')$. The structures of M and M' can be compared with each other in an easy way using the canonical r e d u c t i o n m a p

$$R \ : \ \mathbb{Z} \longrightarrow \mathbb{Z}$$

which we define as follows: Let $m \in \mathbb{Z}$. There is a unique integer $s(m)$ such that $0 \leq s(m) < d$ and $m \equiv s(m)\gamma_0$ modulo d. Then

$$R(m) \ := \ \frac{m - s(m)\gamma_0}{d}.$$

The mapping R is surjective. Each of its fibres consists of d elements. Furthermore, $R(m + dz) = R(m) + z$ for all $z \in \mathbb{Z}$. Because of $dM' \subseteq M$, R maps every fractional ideal of M onto a fractional ideal of M'.

For every $m \in \mathbb{Z}$ the canonical mapping

$$\text{N}_m(\gamma) \longrightarrow \text{N}_{R(m)}(\gamma') \quad \text{by} \quad (\alpha_0, \ldots, \alpha_n) \mapsto \left(\tfrac{\alpha_0 - s(m)}{d}, \alpha_1, \ldots, \alpha_n\right)$$

is bijective. *Proof.* Because of $\alpha_0 = qd + s(m)$ by ordinary Euclidian division, the mapping is well-defined and obviously injective. It is surjective: Let $(\alpha'_0, \ldots, \alpha'_n) \in N_{R(m)}(\gamma')$ be given. Then $m = s(m)\gamma_0 + dR(m) = s(m)\gamma'_0 + d\alpha'_0\gamma'_0 + \cdots + d\alpha'_n\gamma'_n$, such that $(\alpha'_0, \ldots, \alpha'_n)$ is the image of $(\alpha'_0 d + s(m), \alpha'_1, \ldots, \alpha'_n) \in N_m(\gamma)$.

As a consequence, $m \in M$ if and only if $R(m) \in M'$; in particular $M' = R(M)$. Another consequence is, that the canonical isomorphism

$$\Gamma = \mathrm{Syz}(\gamma) \longrightarrow \Gamma' := \mathrm{Syz}(\gamma') \quad \text{by} \quad (a_0, \ldots, a_n) \mapsto (\tfrac{a_0}{d}, a_1, \ldots, a_n)$$

induces an isomorphism $\Gamma(m; \gamma) \to \Gamma(R(m); \gamma')$, for it maps the set $\alpha - \beta$, $\alpha, \beta \in N_m(\gamma)$, of generators of $\Gamma(m; \gamma)$ to the corresponding set of generators of $\Gamma(R(m); \gamma')$. It follows that there is a canonical isomorphism

$$\Gamma/\Gamma(m; \gamma) \overset{\sim}{\longrightarrow} \Gamma'/\Gamma(R(m); \gamma')$$

and that $R(\mathrm{VSat}(\gamma)) = \mathrm{VSat}(\gamma')$, $R(\mathrm{Sat}^r(\gamma)) = \mathrm{Sat}^r(\gamma')$ for all $r \in \mathbb{N}$.

A fractional ideal I of M will be called R-**complete**, if $I = R^{-1}(R(I))$. Explicitly this means: $x = s(x)\gamma_0 + dx'$ belongs to I if and only if all the elements $i\gamma_0 + dx'$, $0 \le i < d$, belong to I. This implies

$$\mathcal{P}_I(Z) = (1 + Z^{\gamma_0} + \cdots + Z^{(d-1)\gamma_0}) \cdot \mathcal{P}_{R(I)}(Z^d)$$
$$= \frac{1 - Z^{d\gamma_0}}{1 - Z^{\gamma_0}} \cdot \mathcal{P}_{R(I)}(Z^d),$$
$$\mathcal{G}_I(Z) = \frac{(1 - Z)(1 - Z^{d\gamma_0})}{(1 - Z^d)(1 - Z^{\gamma_0})} \cdot \mathcal{G}_{R(I)}(Z^d)$$
$$= \frac{1 + Z^{\gamma_0} + \cdots + Z^{(d-1)\gamma_0}}{1 + Z + \cdots + Z^{d-1}} \cdot \mathcal{G}_{R(I)}(Z^d)$$

for the Poincaré series of I and $R(I)$ and their numerators. By straightforward calculations one has the following applications:

3.3 Proposition *Let I be an R-complete fractional ideal of $M = \mathrm{Mon}(\gamma)$ and $R(I)$ the corresponding ideal of $M' = \mathrm{Mon}(\gamma')$. Then*

$$v_I = d\,v_{R(I)}, \qquad g_I = d\,g_{R(I)} + (d-1)\gamma_0, \qquad \vartheta_I = d\,\vartheta_{R(I)} + \frac{(d-1)(\gamma_0 - 1)}{2}.$$

In particular,

$$g_\gamma = d\,g_{\gamma'} + (d-1)\gamma_0, \qquad \vartheta_\gamma = d\,\vartheta_{\gamma'} + \frac{(d-1)(\gamma_0 - 1)}{2},$$
$$\mathrm{sat}^r_\gamma = d\,\mathrm{sat}^r_{\gamma'} + (d-1)\gamma_0, \qquad \mathrm{vsat}_\gamma = d\,\mathrm{vsat}_{\gamma'} + (d-1)\gamma_0.$$

The special cases in 3.3 follow from the fact that the ideals M, $\mathrm{Sat}^r(\gamma)$ and $\mathrm{VSat}(\gamma)$ are R-complete, a consequence of $\Gamma/\Gamma(m; \gamma) \cong \Gamma'/\Gamma(R(m); \gamma')$ for all $m \in \mathbb{Z}$.

Supplements

1. Let M be a numerical monoid.

a) The following conditions are equivalent:
(1) M is symmetric, i.e. $t_M = 1$.
(2)   For every $x \in \mathbb{Z}$, $x \notin M$, there is an element $u \in M$ with $x + u = g_M$.
(3)   For every $x \in \mathbb{Z}$ either x or $g_M - x$ belongs to M.
(4)   The following mapping is bijective:
$$M \cap \{0, \ldots, g_M\} \longrightarrow \mathbb{N} \smallsetminus M \quad \text{by} \quad x \mapsto g_M - x.$$
(This motivates the notation "symmetric".)

b) $g_M \leq 2\vartheta_M - 1$. Equality holds if and only if M is symmetric.

2. (J.J. Sylvester) Let $\gamma_0, \gamma_1 \in \mathbb{N}^*$ be relatively prime and let $M := \mathrm{Mon}(\gamma_0, \gamma_1)$. Then $g_M = \gamma_0\gamma_1 - \gamma_0 - \gamma_1$ and M is symmetric. (For every $x \in \mathbb{Z}$ there are integers a, b such that $x = a\gamma_0 + b\gamma_1$ with $0 \leq a \leq \gamma_1 - 1$.)

3. Let $\gamma = (\gamma_0, \gamma_1, \gamma_2)$ with $\gcd(\gamma_0, \gamma_1, \gamma_2) = 1$ and let $K := \mathrm{lcm}(\gamma_0, \gamma_1, \gamma_2)$, $d := \gcd(\gamma_0, \gamma_1)$. For every $m \in \mathbb{N}^*$ the following holds: dm or $K - dm$ belongs to $\mathrm{Mon}(\gamma_0, \gamma_1)$. (Reduce to the case $d = 1$.)

4. Let M be a numerical monoid and I a fractional ideal of M.

a) $(\vartheta_{I^{\smallfrown}} + \mathrm{v}_{I^{\smallfrown}}) + (\vartheta_I + \mathrm{v}_I) = 1$. (Duality theorem)
(Multiplication by -1 maps $\{z : z \geq \mathrm{v}_{I^{\smallfrown}}, z \notin I^{\smallfrown}\} = \{z : z \geq -g_I, -z \in I\}$ onto $\{u : u \in I, \mathrm{v}_I \leq u \leq g_I\}$.)

b) $\vartheta_{\Omega_M} \leq \vartheta_M$. Equality holds if and only if M is symmetric. (Follows from the inclusion $-g_M + M \subseteq \Omega_M$. The inequality $\vartheta_I \leq \vartheta_M$ holds for every fractional ideal I.)

5. Let M be a numerical monoid. The following conditions are equivalent:
(1) M is symmetric.
(2) $\Omega_M = -g_M + M$.   (2') $\mathcal{P}_{\Omega_M} = Z^{-g_M}\mathcal{P}_M$.
(3) $\mathcal{P}_M = -Z^{g_M}\mathcal{P}_M(1/Z)$.
(4) $\mathcal{G}_M = Z^{g_M+1}\mathcal{G}_M(1/Z)$, i.e. $\mathcal{G}_M$ is a self-reciprocal polynomial.

6. (*Arithmetic progressions.* A. Brauer, E.S. Selmer et al.) Let a, b, n be positive natural numbers with $\gcd(a, b) = 1$. The sequence
$$\gamma_i := a + ib, \quad i = 0, \ldots, n,$$
generates a numerical monoid $M := \mathrm{Mon}(\gamma_0, \ldots, \gamma_n)$. Let $a - 2 = qn + r$, $q \in \mathbb{Z}$, $0 \leq r < n$. Then
$$g_M = (a-1)(b+q) + q, \quad 2\vartheta_M - 1 = g_M + r(q+1).$$
($\gamma_i + j\gamma_n = (1+j)a + (i+jn)b$ for $1 \leq i \leq n$, $0 \leq j < q$ and $1 \leq i \leq r+1$, $j = q$ are the least elements of M to cover all residue classes $\neq 0$ modulo a, i.e. the elements of the standard basis of M with respect to a. Let S denote their sum. Then $g_M = \gamma_{r+1} + q\gamma_n - a$ and $\vartheta_M = S/a - (a-1)/2 = ((a-1)(b+q) + (r+1)(q+1))/2$.)

7. In the situation of 3.1 let $d_i := \gcd(\gamma_j; j \neq i)$. Then the first inequality of 3.1(2) can be improved to $g_\gamma + \max_i d_i\gamma_i \leq \mathrm{sat}_\gamma$.

8. Let $\gamma = (\gamma_0, \ldots, \gamma_n)$ with $\gcd(\gamma_0, \ldots, \gamma_n) = 1$. Then
$$g_\gamma \leq \mathrm{lcm}(\gamma_0, \gamma_1) + \cdots + \mathrm{lcm}(\gamma_0, \gamma_n) - |\gamma|.$$

(For $m \in \mathbb{Z}$ there is a representation $m = a_0\gamma_0 - a_1\gamma_1 - \cdots - a_n\gamma_n$ over $\mathbb{Z}$ with $0 \leq a_i\gamma_i < \mathrm{lcm}(\gamma_0, \gamma_i)$ for $i = 1, \ldots, n$.)

9. Let $\gamma = (\gamma_0, \ldots, \gamma_n)$ with $\gcd(\gamma_0, \ldots, \gamma_n) = 1$ and $\gamma_0 \leq \gamma_1 \leq \cdots \leq \gamma_n$. Then
$$g_\gamma \leq 2\vartheta_\gamma - 1 \leq \gamma_0\gamma_n - (\gamma_0 + \gamma_n), \quad \mathrm{sat}_\gamma \leq \mathrm{vsat}_\gamma \leq 2\gamma_0\gamma_n - (\gamma_0 + \gamma_n).$$
(The inequalities not yet known are proved by induction on n, reducing first to the case that $\gcd(\gamma_0, \ldots, \gamma_{n-1}) = 1$.)

10. Let $\gamma = (\gamma_0, \ldots, \gamma_n)$ with $\gcd(\gamma_0, \ldots, \gamma_n) = 1$, $\bar{\gamma} := (\gamma_0, \ldots, \gamma_n, \gamma_{n+1})$ with any $\gamma_{n+1} \in \mathbb{N}^*$ and $M := \mathrm{Mon}(\gamma)$. Then
$$g_{\bar{\gamma}} \leq g_\gamma, \quad \vartheta_{\bar{\gamma}} \leq \vartheta_\gamma.$$
Furthermore, $\mathrm{Sat}(\bar{\gamma}) \supseteq \mathrm{Sat}(\gamma) \cap (M + \gamma_{n+1})$ and $\mathrm{VSat}(\bar{\gamma}) \supseteq \mathrm{VSat}(\gamma) \cap (M + \gamma_{n+1})$, such that
$$\mathrm{sat}_{\bar{\gamma}} \leq \max(\mathrm{sat}_\gamma, g_\gamma + \gamma_{n+1}), \quad \mathrm{vsat}_{\bar{\gamma}} \leq \max(\mathrm{vsat}_\gamma, g_\gamma + \gamma_{n+1}).$$
(The homomorphism h from the proof of 3.2 is for every $m \in M \cap (M + \gamma_{n+1})$ surjective.)

11. Let the numerical monoid M be related to a numerical monoid M' by the process of reduction in codimension 1. Then M is symmetric if and only if M' is symmetric.

12. a) Let $\gamma = (\gamma_0, \gamma_1)$ be a pair of relatively prime positive integers. Then $g_\gamma = \gamma_0\gamma_1 - (\gamma_0 + \gamma_1)$, $\mathrm{sat}_\gamma = \mathrm{vsat}_\gamma = g_\gamma + \gamma_0\gamma_1 = 2\gamma_0\gamma_1 - (\gamma_0 + \gamma_1)$. (See also Suppl. 2.)

b) More generally, for arbitrary $n \geq 1$, let $a_0, \ldots, a_n$ be pairwise relatively prime positive integers, $b := a_0 \cdots a_n$ their product and $\gamma := (\gamma_0, \ldots, \gamma_n)$ with $\gamma_i := b/a_i$. Then
$$g_\gamma = nb - |\gamma|, \quad \mathrm{sat}_\gamma = \mathrm{vsat}_\gamma = g_\gamma + b$$
and $g_\gamma = 2\vartheta_\gamma - 1$, i.e. $\mathrm{Mon}(\gamma)$ is symmetric. (Reduction in codimension 1. Remark: $\mathrm{Mon}(\gamma)$ is even a complete intersection, cf. Suppl. 7 in Sect. 4.)

13.† Let $\gamma = (\gamma_0, \ldots, \gamma_n)$, $n \geq 1$, generate the numerical monoid $M = \mathrm{Mon}(\gamma)$, let $d \in \mathbb{N}^*$ be a common divisor of $\gamma_1, \ldots, \gamma_n$, finally $\gamma_0' := \gamma_0$, $\gamma_i' := \gamma_i/d$ for $i = 1, \ldots, n$, $\gamma' := (\gamma_0', \ldots, \gamma_n')$ and $M' := \mathrm{Mon}(\gamma')$. Beside the reduction map R there is another (canonical) map $R^\wedge : \mathbb{Z} \to \mathbb{Z}$ which is defined as follows: Let $m \in \mathbb{Z}$. There is a unique integer $\hat{s}(m)$ such that $0 \leq \hat{s}(m) < d$ and $m \equiv -\hat{s}(m)\gamma_0$ modulo d. Then
$$R^\wedge(m) := \frac{m + \hat{s}(m)\gamma_0}{d}.$$
$R^\wedge$ is surjective. $R^\wedge(m + dz) = R^\wedge(m) + z$ for all $z \in \mathbb{Z}$. Thus $R^\wedge$ maps every fractional ideal of M onto a fractional ideal of M'.

a) $R^\wedge(-z) = -R(z)$ for all $z \in \mathbb{Z}$.

b) A fractional ideal I of M is called $R^\wedge$-c o m p l e t e if $I = (R^\wedge)^{-1}(R^\wedge(I))$. Explicitly this means: $x = -\hat{s}(x)\gamma_0 + dx'$ belongs to I if and only if all the elements $-i\gamma_0 + dx'$, $0 \leq i < d$, belong to I. If $d > 1$, M is not $R^\wedge$-complete. The ideal Ω_M is always $R^\wedge$-complete. In addition:
  I is R-complete if and only if $I^\wedge$ is $R^\wedge$-complete. In this case $R^\wedge(I^\wedge) = R(I)^\wedge$.

c) Let I be a fractional ideal of M. If I is R-complete, then R maps the minimal set of generators of I bijectively onto the minimal set of generators of $R(I)$. An analogous result holds for $R^\wedge : I \to R^\wedge(I)$, if I is $R^\wedge$-complete. There are natural applications to the type sets and the types of monoids.

d) Let I be a fractional M-ideal. If I is R-complete,
$$\mathrm{v}_I \;=\; d\,\mathrm{v}_{R(I)} \,.$$
If I is $R^{\smallfrown}$-complete,
$$\mathrm{v}_I \;=\; d\,\mathrm{v}_{R^{\smallfrown}(I)} - (d-1)\gamma_0 \,.$$
e) For every $R^{\smallfrown}$-complete fractional ideal I of M the following holds:
$$\mathrm{g}_I \;=\; d\,\mathrm{g}_{R^{\smallfrown}(I)}, \qquad \vartheta_I \;=\; d\,\vartheta_{R^{\smallfrown}(I)} + \tfrac{(d-1)(\gamma_0-1)}{2} \,.$$

14.† Let $\gamma = (\gamma_0,\dots,\gamma_n)$ generate the numerical monoid $M = \mathrm{Mon}(\gamma)$.

a) If $n \geq 1$, then, for every $m \in \mathrm{Sat}^{n-1}(\gamma)$, the group $\Gamma(m;\gamma)$ contains a direct summand of rank 1 of $\mathrm{Syz}(\gamma)$.
(Let $m \in \mathrm{Sat}^{n-1}(\gamma)$. There are two different representations $m = \langle \alpha, \gamma \rangle = \langle \beta, \gamma \rangle$. If $t := \gcd(\beta - \alpha)$ then $\beta' := \alpha + \frac{1}{t}(\beta - \alpha) = (1 - \frac{1}{t})\alpha + \frac{1}{t}\beta$ is an element of $\mathrm{N}_m(\gamma)$ and $\beta' - \alpha$ belongs to a basis of $\mathrm{Syz}(\gamma)$.)

b) If $n \geq 2$, then, for every $m \in \mathrm{Sat}^{n-2}(\gamma)$, the group $\Gamma(m;\gamma)$ contains a direct summand of rank 2 of $\mathrm{Syz}(\gamma)$.
(One uses the following *Lemma*: Let $\alpha(0), \alpha(1), \alpha(2) \in \mathrm{N}_m(\gamma)$ be such that $x := \alpha(1) - \alpha(0)$, $y := \alpha(2) - \alpha(0)$ are linearly independent. Then x, y is part of a $\mathbb{Z}$-basis of $\mathrm{Syz}(\gamma)$ if and only if the closed triangle S with vertices $\alpha(0), \alpha(1), \alpha(2)$ in $\mathbb{Q}^{n+1}$ contains no other points of $\mathbb{Z}^{n+1}$ but $\alpha(0), \alpha(1), \alpha(2)$. *Proof.* Using the reflection $z \mapsto -z + \alpha(1) + \alpha(2)$ the condition on S is seen to be equivalent with the condition that the parallelogram with vertices $\alpha(0), \alpha(1), \alpha(2), \alpha(1) + \alpha(2) - \alpha(0)$ contains no other points of $\mathbb{Z}^{n+1}$ but its vertices. This is equivalent to the fact that x, y form a basis of the two-dimensional lattice $(\mathbb{Q}\,x + \mathbb{Q}\,y) \cap \mathbb{Z}^{n+1}$.

Let $m \in \mathrm{Sat}^{n-2}(\gamma)$. Choose $\alpha(0), \alpha(1), \alpha(2) \in \mathrm{N}_m(\gamma)$ in such a way that $\alpha(1) - \alpha(0)$, $\alpha(2) - \alpha(0)$ are linearly independent and that the triangle S with vertices $\alpha(0), \alpha(1), \alpha(2)$ has minimal volume. Then S contains no other points of $\mathrm{N}_m(\gamma) = \mathbb{N}^{n+1} \cap \{z \in \mathbb{Q}^{n+1} : \langle z, \gamma \rangle = m\}$ and therefore no other points of $\mathbb{Z}^{n+1}$ but $\alpha(0), \alpha(1), \alpha(2)$. Now apply the Lemma.)

In particular, if M can be generated by ≤ 3 elements, then $\mathrm{Sat}(\gamma) = \mathrm{VSat}(\gamma)$ and $\mathrm{sat}_\gamma = \mathrm{vsat}_\gamma$.

4 Relations of Numerical Monoids

Let $M = \text{Mon}(\gamma)$ be a numerical monoid $(\subseteq \mathbb{N})$ with a given system $\gamma = (\gamma_0, \ldots, \gamma_n)$ of positive generators. For every commutative ring A the canonical surjective monoid homomorphism

$$\pi^\gamma : \mathbb{N}^{n+1} \longrightarrow M$$

with $\alpha \mapsto \langle \alpha, \gamma \rangle$ gives rise to a homogeneous A-algebra homomorphism

$$\pi_A^\gamma : A[T] = A[T_0, \ldots, T_n] \longrightarrow A[M] \subseteq A[\mathbb{N}] = A[X]$$

which maps the indeterminate T_i of degree γ_i to $\gamma_i = X^{\gamma_i}$, $i = 0, \ldots, n$. The image of π_A^γ is $A[M]$. Its kernel

$$\mathfrak{I} := \ker \pi_A^\gamma$$

is called the ideal of relations of M with respect to the representation $M = \text{Mon}(\gamma)$.

A relation of M itself (with respect to γ) is a pair $(\alpha, \beta) \in \mathbb{N}^{n+1} \times \mathbb{N}^{n+1}$ with $\langle \alpha, \gamma \rangle = \langle \beta, \gamma \rangle$. To such a relation of M there corresponds the binomial $T^\alpha - T^\beta$ in the ideal of relations $\mathfrak{I}$. *The binomials of this type generate the ideal $\mathfrak{I}$. Proof.* Because $\mathfrak{I}$ is homogeneous let us consider a homogeneous polynomial $F = \sum_k a_k T^{\alpha(k)} \in \mathfrak{I}$ of degree $m \geq 0$, $a_k \in A$, $\alpha(k) \in \mathbb{N}_m(\gamma)$. From $0 = \pi_A^\gamma(F) = (\sum_k a_k) X^m$ we get $\sum_k a_k = 0$. Therefore, for a fixed index κ, $F = \sum_k a_k (T^{\alpha(k)} - T^{\alpha(\kappa)})$.

The following remark is a simple application of the last result (in case $A \neq 0$): The M-ideal $\{m \in M : \mathfrak{I}_m \neq 0\}$ coincides with the M-ideal $\text{Sat}^{n-1}(\gamma)$ introduced in Section 3, and the degrees of any set of homogeneous generators of $\mathfrak{I}$ generate $\text{Sat}^{n-1}(\gamma)$.

The set of all relations of M is a compatible equivalence relation on $\mathbb{N}^{n+1}$. In general, an equivalence relation R on the monoid $\mathbb{N}^{n+1}$ is called compatible if $(\alpha, \beta), (\alpha', \beta') \in R$ implies $(\alpha + \alpha', \beta + \beta') \in R$. For a compatible equivalence relation R the quotient $\mathbb{N}^{n+1}/R$ carries a unique monoid structure such that the canonical projection $\mathbb{N}^{n+1} \to \mathbb{N}^{n+1}/R$ is a monoid homomorphism.

Let $(\alpha(j), \beta(j))$, $j \in J$, be any system of relations of M. This system generates a smallest compatible equivalence relation R containing all the $(\alpha(j), \beta(j))$ and π^γ induces a surjective monoid homomorphism $\mathbb{N}^{n+1}/R \to M$, which is an isomorphism if and only if R is the set of all relations of M. In this case $(\alpha(j), \beta(j))$, $j \in J$, is called a system of generators of all relations of M.

4.1 Proposition *Let $(\alpha(j), \beta(j))$, $j \in J$, be a system of relations of M and let R be the smallest compatible equivalence relation on $\mathbb{N}^{n+1}$ containing all the $(\alpha(j), \beta(j))$. Then the kernel $\mathfrak{A}$ of the surjective algebra homomorphism*

$$A[T] \longrightarrow A[\mathbb{N}^{n+1}/R]$$

induced by the canonical projection $\mathbb{N}^{n+1} \to \mathbb{N}^{n+1}/R$ is generated by the binomials $T^{\alpha(j)} - T^{\beta(j)}$, $j \in J$.

Proof. Let $\mathfrak{A}' \subseteq \mathfrak{A}$ be the ideal generated by the binomials $T^{\alpha(j)} - T^{\beta(j)}$ and for $\alpha \in \mathbb{N}^{n+1}$ let t^α denote the residue class of T^α in $A[T]/\mathfrak{A}'$. Then the monoid homomorphism $\alpha \mapsto t^\alpha$, $\alpha \in \mathbb{N}^{n+1}$, induces a monoid homomorphism on $\mathbb{N}^{n+1}/R$ which defines an algebra homomorphism $A[\mathbb{N}^{n+1}/R] \to A[T]/\mathfrak{A}'$. The composition with $A[T] \to A[\mathbb{N}^{n+1}/R]$ is the canonical projection $A[T] \to A[T]/\mathfrak{A}'$ which implies that $\mathfrak{A}$ is mapped to zero in $A[T]/\mathfrak{A}'$, i. e. $\mathfrak{A} \subseteq \mathfrak{A}'$. $\square$

4.2 Corollary *Let $A \neq 0$. A system $(\alpha(j), \beta(j))$, $j \in J$, of relations of M generates all the relations of M if and only if the corresponding homogeneous binomials $T^{\alpha(j)} - T^{\beta(j)}$, $j \in J$, generate the ideal $\mathfrak{I} = \ker \pi_A^\gamma \subseteq A[T]$.*

Note that the criterion in 4.2 is independent of the coefficient ring $A \neq 0$. Because for a noetherian ring A the ideal $\mathfrak{I}$ is finitely generated, this is true for an arbitrary A. Therefore, any minimal set of generators of $\mathfrak{I}$ is finite.

The minimal number $\mu(\mathfrak{I}) = \mu_{A[T]}(\mathfrak{I})$ of generators of $\mathfrak{I}$ is independent of the ring A $(\neq 0)$. Every minimal set of binomials generating $\mathfrak{I}$ contains exactly $\mu(\mathfrak{I})$ elements. Proof. Let F_j, $j \in J$, be a minimal system of generators of $\mathfrak{I} \subseteq A[T]$. Then this system generates the kernel $\bar{\mathfrak{I}}$ of π_K^γ for any residue class field K of A. card J is greater than or equal to the dimension $\dim_K \bar{\mathfrak{I}}/\mathfrak{M}\bar{\mathfrak{I}}$, $\mathfrak{M} := K[T]_+$. By the Lemma of Krull-Nakayama for graded K-algebras, any minimal system of homogeneous generators of $\bar{\mathfrak{I}}$ contains exactly $\dim_K \bar{\mathfrak{I}}/\mathfrak{M}\bar{\mathfrak{I}}$ elements. Now assume that the F_j are homogeneous binomials. There is a subset $J' \subseteq J$ such that F_j, $j \in J'$, generate $\bar{\mathfrak{I}}$ minimally. But $J' = J$ by 4.2. Thus card $J = \dim_K \bar{\mathfrak{I}}/\mathfrak{M}\bar{\mathfrak{I}} = \mu(\mathfrak{I})$ in this case.

Let $A = K$ be a field. Then

$$\mu(\mathfrak{I}) \;=\; \dim_K \mathfrak{I}/\mathfrak{M}\mathfrak{I} \;=\; \mathcal{R}_\gamma(1)$$

where

$$\mathcal{R}_\gamma := \sum_{m \in \mathbb{N}} d_m(\gamma) Z^m \,,$$

$d_m(\gamma) := \dim_K (\mathfrak{I}/\mathfrak{M}\mathfrak{I})_m$, is the Poincaré series of $\mathfrak{I}/\mathfrak{M}\mathfrak{I}$, which is actually a polynomial. (As the proof above shows, also the polynomial $\mathcal{R}_\gamma$ is independent of the field K. Even for an arbitrary ring $A \neq 0$, $d_m(\gamma)$ is the rank of the free A-module $(\mathfrak{I}/\mathfrak{M}\mathfrak{I})_m$, $\mathfrak{I} = \ker \pi_A^\gamma$, $\mathfrak{M} = A[T]_+$. For a purely combinatorial description of $\mathcal{R}_\gamma$ see the remarks following Proposition 4.3 below.)

A system $(\alpha(j), \beta(j))$, $j \in J$, of relations of $M = \text{Mon}(\gamma)$ will be called a m i n i m a l   s y s t e m   o f   r e l a t i o n s, if the binomials $T^{\alpha(j)} - T^{\beta(j)}$, $j \in J$, form a minimal system of generators of $\mathfrak{I}$.

The number

$$\text{dev}_M = \text{dev}_\gamma := \mu(\mathfrak{I}) - n = \mathcal{R}_\gamma(1) - n$$

is independent of the representation $M = \text{Mon}(\gamma)$ of M because for any field K the number $\mu(\mathfrak{I}) - n = \mu(\mathfrak{I}) - (n+1) + 1$ is the deviation ("Abweichung") of the one-dimensional local ring $K[\![M]\!] = K[\![T]\!]/\mathfrak{I}K[\![T]\!]$ which is the completion of $K[M]$ with respect to its homogeneous maximal ideal. (For the deviation of a noetherian local ring cf. Section 6, Supplement 5.) In the situation here an elementary proof is possible, see Supplement 4 a). We will call dev_M the d e v i a t i o n of the numerical monoid M.

For simple reasons $\text{dev}_M \geq 0$, see Supplement 1 or use the following argument: If $(\alpha(j), \beta(j))$, $j \in J$, is a system of generators of all the relations of M, then the differences $\alpha(j) - \beta(j)$, $j \in J$, generate the group $\text{Syz}(\gamma)$ of syzygies of γ, which is a group of rank n. Thus $|J| \geq n$.

In case $\text{dev}_M = 0$, the monoid M is called a c o m p l e t e   i n t e r s e c t i o n. We mention that M is a complete intersection if and only if for every noetherian ring $A \neq 0$ the monoid algebra $A[M]$ is a graded complete intersection over A. For details see Supplement 2.

As our next topic we propose to study where the minimal binomial generators of the ideal $\mathfrak{I} = \ker \pi_A^\gamma$ are situated. For this let again $A = K$ be a field. Then we have to determine bases of the K-vector spaces $(\mathfrak{I}/\mathfrak{M}\mathfrak{I})_m = \mathfrak{I}_m/(\mathfrak{M}\mathfrak{I})_m$, $m \in \mathbb{N}$, $\mathfrak{M} := K[T]_+$. It is more convenient to study the spaces $K[T]_m/(\mathfrak{M}\mathfrak{I})_m$ and look out for their monomial bases.

$(\mathfrak{M}\mathfrak{I})_m$ is generated over K by the binomials $T^\alpha - T^\beta$ with $\alpha, \beta \in \mathbb{N}_m(\gamma)$ and $\gcd(T^\alpha, T^\beta) \neq 1$, i.e. $\text{supp}(\alpha) \cap \text{supp}(\beta) \neq \emptyset$. Here the s u p p o r t of any $\alpha \in \mathbb{N}^{n+1}$ is defined to be the set

$$\text{supp}(\alpha) := \{i : 0 \leq i \leq n, \alpha_i > 0\} \subseteq \{0, \ldots, n\}.$$

This description of the generators of $(\mathfrak{M}\mathfrak{I})_m$ suggests to introduce the following graph structure on the set $\mathbb{N}_m(\gamma)$: Two points $\alpha, \beta \in \mathbb{N}_m(\gamma)$, $\alpha \neq \beta$, are joined by an edge if $\text{supp}(\alpha) \cap \text{supp}(\beta) \neq \emptyset$. Let the set of the connected components be denoted by

$$\bar{\mathbb{N}}_m(\gamma)$$

and its cardinal number by

$$c_m = c_m(\gamma).$$

A polynomial $\sum_{\alpha \in N_m(\gamma)} a_\alpha T^\alpha \in K[T]_m$ *belongs to* $(\mathfrak{M}\mathfrak{I})_m$ *if and only if* $\sum_{\alpha \in S} a_\alpha = 0$ *for every connected component* $S \in \bar{N}_m(\gamma)$. This is a well-known and simple result of graph theory. The *proof* runs as follows: The binomial generators $T^\alpha - T^\beta$ of $(\mathfrak{M}\mathfrak{I})_m$ fulfil these relations, so do all elements of $(\mathfrak{M}\mathfrak{I})_m$. For the converse it is sufficient to prove that any polynomial $F = \sum_{\alpha \in S} a_\alpha T^\alpha$ with $\sum_{\alpha \in S} a_\alpha = 0$ belongs to $(\mathfrak{M}\mathfrak{I})_m$. We fix an element $\beta \in S$. Then $F = \sum_{\alpha \in S} a_\alpha (T^\alpha - T^\beta)$. It suffices to show that $T^\alpha - T^\beta \in (\mathfrak{M}\mathfrak{I})_m$. There is a chain $\beta = \alpha(0), \alpha(1), \ldots, \alpha(r) = \alpha$ such that $\alpha(i) \in N_m(\gamma)$ and $\text{supp}\,\alpha(i) \cap \text{supp}\,\alpha(i+1) \neq \emptyset$ for $i = 0, \ldots, r-1$. Then $T^\alpha - T^\beta = \sum_{i=0}^{r-1} (T^{\alpha(i+1)} - T^{\alpha(i)}) \in (\mathfrak{M}\mathfrak{I})_m$.

A monomial basis of $K[T]_m/(\mathfrak{M}\mathfrak{I})_m$ is given by (the residue classes of) $T^{\alpha(j)}$, where $\alpha(j) \in N_m(\gamma)$, $j = 1, \ldots, c_m(\gamma)$, is a full system of representatives for $\bar{N}_m(\gamma)$. Therefore we have:

4.3 Proposition *Let* $m \in \mathbb{N}$. *Then* $\mathfrak{I}_m \neq (\mathfrak{M}\mathfrak{I})_m$ *if and only if* $c_m(\gamma) > 1$. *In this case a basis of* $\mathfrak{I}_m$ *modulo* $(\mathfrak{M}\mathfrak{I})_m$ *can be given by the binomials*

$$T^{\alpha(j)} - T^{\alpha(1)}, \quad j = 2, \ldots, c_m(\gamma),$$

where $\alpha(1), \ldots, \alpha(c_m(\gamma))$ *is a full system of representatives for the set* $\bar{N}_m(\gamma)$ *of the connected components of* $N_m(\gamma)$.

A similar description of $\mathfrak{I}_m/(\mathfrak{M}\mathfrak{I})_m$ is given in [7].

The connected components of $N_m(\gamma)$, $m > 0$, can be described in a simpler way. To do this, define

$$G_m(\gamma) := \bigcup_{\alpha \in N_m(\gamma)} \text{supp}(\alpha).$$

Then the sets $\bigcup_{\alpha \in S} \text{supp}(\alpha)$, $S \in \bar{N}_m(\gamma)$, obviously form a partition of $G_m(\gamma)$ into non-empty subsets of $G_m(\gamma)$. Let

$$\bar{G}_m(\gamma)$$

denote the set of elements of this partition. Then $\text{card}\,\bar{G}_m(\gamma) = c_m(\gamma)$, and the elements of $\bar{G}_m(\gamma)$ are the connected components of the following graph on the set $G_m(\gamma)$ as set of vertices: Two points $r, s \in G_m(\gamma)$, $r \neq s$, are joined by an edge if and only if there is an $\alpha \in N_m(\gamma)$ with $r, s \in \text{supp}(\alpha)$. The simple proof is left to the reader.

The set of $m \in \mathbb{N}$ such that $G_m(\gamma) = \{0, \ldots, n\}$ and $c_m(\gamma) = 1$ is a subset of M and even a non-empty ideal of M, which we shall denote by

$$\text{Con}(\gamma).$$

Its Frobenius number we denote by con_γ. Then

$$\text{con}_\gamma \leq g_\gamma + |\gamma|$$

because for $m > g_\gamma + \gamma_0 + \cdots + \gamma_n$ there is an $\alpha \in N_m(\gamma)$ with $\mathrm{supp}(\alpha) = \{0, \ldots, n\}$. Note that every $m \in M$, which is the degree of an element of a minimal system of homogeneous generators of $\mathfrak{I}$, does not belong to $\mathrm{Con}(\gamma)$; in particular, $\mathrm{rmax}_\gamma \leq \mathrm{con}_\gamma$, where

$$\mathrm{rmax}_\gamma = \deg \mathcal{R}_\gamma$$

denotes the supremum of the degrees of the elements of such a system. See Supplements 5, 6 below and Supplement 1 in Section 5 for further details. The coefficients of the Poincaré series $\mathcal{R}_\gamma = \sum_{m \in \mathbb{N}} d_m(\gamma) Z^m$ of $\mathfrak{I}/\mathfrak{M}\mathfrak{I}$ can be described as $d_m(\gamma) = c_m(\gamma) - 1$, if $c_m(\gamma) > 0$, and $d_m(\gamma) = 0$, otherwise. Recall that the equality

$$\mathcal{R}_\gamma(1) = \sum_{m \in \mathbb{N}} d_m(\gamma) = n$$

characterizes the monoids which are complete intersections.

Let A be again an arbitrary commutative ring. The monoid homomorphism $\pi^\gamma : \mathbb{N}^{n+1} \to M$ is the restriction of the homomorphism $(\pi^\gamma) : \mathbb{Z}^{n+1} \to \mathbb{Z}$ with $\alpha \mapsto \langle \alpha, \gamma \rangle$. This group homomorphism gives rise to a homogeneous A-algebra homomorphism

$$(\pi_A^\gamma) \; : \; A[T^{\pm 1}] = A[T_0^{\pm 1}, \ldots, T_n^{\pm 1}] \longrightarrow A[\mathbb{Z}] = A[X^{\pm 1}]$$

from the graded algebra $A[T^{\pm 1}] = A[T]_{T_0 \cdots T_n}$ of Laurent polynomials in the indeterminates $T_0, \ldots, T_n$ onto the graded algebra $A[X^{\pm 1}] = A[X]_X$ of Laurent polynomials in X. The homomorphism $(\pi_A^\gamma) = (\pi_A^\gamma)_{T_0 \cdots T_n}$ is obtained from π_A^γ simply by forming rings of fractions (Nenneraufnahme) with respect to multiplicative sets generated by $T_0, \ldots, T_n$ and X, respectively. In particular,

$$\ker(\pi_A^\gamma) = \mathfrak{I}A[T^{\pm 1}] \quad \text{and} \quad \mathfrak{I} = A[T] \cap (\mathfrak{I}A[T^{\pm 1}]).$$

Note that for every relation (α, β) of the monoid M the binomial

$$T^\alpha - T^\beta = T^\beta(T^{\alpha - \beta} - 1) \in A[T^{\pm 1}]$$

is associated to the Laurent polynomial $T^s - 1$, where $s := \alpha - \beta \in \mathrm{Syz}(\gamma) = \ker(\pi^\gamma)$.

4.4 Theorem *Assume $A \neq 0$.*

(1) *A system $s(p) \in \ker(\pi^\gamma) = \mathrm{Syz}(\gamma)$, $p \in P$, of syzygies of γ generates $\mathrm{Syz}(\gamma)$ if and only if the Laurent polynomials $T^{s(p)} - 1$, $p \in P$, generate the ideal $\ker(\pi_A^\gamma) = \mathfrak{I}A[T^{\pm 1}]$.*

(2) *Let $(\alpha(j), \beta(j))$, $j \in J$, be a minimal system of relations of M, $\delta_j := \langle \alpha(j), \gamma \rangle = \langle \beta(j), \gamma \rangle$ the degrees of the corresponding binomials $F_j := T^{\alpha(j)} - T^{\beta(j)} \in \mathfrak{I}$ and let $m \in \mathbb{N}$. Then the ideal*

$$\mathfrak{I}_m A[T^{\pm 1}] = \sum_{j \text{ with } m \in M + \delta_j} F_j A[T^{\pm 1}]$$

is generated by $T^s - 1$, $s \in \Gamma(m; \gamma)$. Thus

$$\Gamma(m; \gamma) = \sum_{j \text{ with } m \in M + \delta_j} \mathbb{Z} \cdot (\alpha(j) - \beta(j)).$$

Proof. For any system $s(p) \in \mathrm{Syz}(\gamma)$, $p \in P$, $A[T^{\pm 1}]/(T^{s(p)} - 1; p \in P)$ is canonically isomorphic to the group algebra $A[\mathbb{Z}^{n+1} / \sum_{p \in P} \mathbb{Z} s(p)]$ over A. From this (1) follows directly. The chain

$$\sum_{j \text{ with } m \in M + \delta_j} F_j A[T^{\pm 1}] \subseteq \sum_{s \in \Gamma(m; \gamma)} (T^s - 1) A[T^{\pm 1}]$$

$$\subseteq \mathfrak{I}_m A[T^{\pm 1}] \subseteq \sum_{j \text{ with } m \in M + \delta_j} F_j A[T^{\pm 1}]$$

of inclusions yields (2). $\qquad\qquad\qquad\qquad\qquad\qquad\qquad\qquad\qquad\square$

4.5 Corollary *In the situation of Theorem 4.4(2), for $k \in \mathbb{N}$ let $E^k(J)$ denote the set of subsets $E \subseteq J$ such that $|E| = k$ and that the syzygies $\alpha(j) - \beta(j)$, $j \in E$, are linearly independent. Then, for arbitrary $r \in \mathbb{N}$,*

$$\mathrm{Sat}^r(\gamma) = \bigcup_{E \in E^{n-r}(J)} \Big(\bigcap_{j \in E} M + \delta_j \Big).$$

Proof. $m \in \mathrm{Sat}^r(\gamma)$ if and only if $\mathrm{rank}\,\Gamma(m; \gamma) \geq n - r$. Therefore the assertion follows by 4.4(2). $\qquad\qquad\qquad\qquad\qquad\qquad\qquad\qquad\qquad\square$

4.6 Corollary *In the situation of Theorem 4.4 let the monoid $M = \mathrm{Mon}(\gamma)$ be a complete intersection. Then $\Gamma(m; \gamma)$ is a direct summand of $\mathrm{Syz}(\gamma)$ for every $m \in \mathbb{N}$, and for every $r \in \mathbb{N}$*

$$\mathrm{Sat}^r(\gamma) = \bigcup_{E \subseteq J, |E| = n-r} \Big(\bigcap_{j \in E} M + \delta_j \Big).$$

Proof. If M is a complete intersection, then $\alpha(j) - \beta(j)$, $j \in J$, is a basis of $\mathrm{Syz}(\gamma)$. So the result follows from Corollary 4.5. $\qquad\qquad\qquad\square$

Supplements

1. a) Let K be a field, $F_1, \ldots, F_m$ arbitrary homogeneous polynomials of positive degrees $\delta_1, \ldots, \delta_m$ in the graded polynomial algebra $K[T_0, \ldots, T_n]$, $\deg T_i = \gamma_i > 0$, $i = 0, \ldots, n$, and B the algebra $K[T_0, \ldots, T_n]/(F_1, \ldots, F_m)$. Then

$$\frac{1}{(1 - Z)^m} \cdot \mathcal{P}_B \geq \frac{1}{(1 - Z)^m} \cdot \frac{(1 - Z^{\delta_1}) \cdots (1 - Z^{\delta_m})}{(1 - Z^{\gamma_0}) \cdots (1 - Z^{\gamma_n})},$$

i. e. every coefficient of the power series on the left side is greater than or equal to the corresponding coefficient of the power series on the right side. (Induction on m.)

b) For a numerical monoid $M = \text{Mon}(\gamma_0, \ldots, \gamma_n)$, the ideal $\mathfrak{J} = \ker \pi_K^\gamma \subseteq K[T_0, \ldots, T_n]$ is generated by at least n elements. (Let $F_1, \ldots, F_m$ be homogeneous generators of the ideal $\mathfrak{J}$. By a) now, $1/(1-Z)^{m+1} = \mathcal{P}_{K[X]}/(1-Z)^m \geq \mathcal{P}_{K[M]}/(1-Z)^m \geq 1/(1-Z^{\gamma_0})\cdots(1-Z^{\gamma_n})$. This is impossible for $m < n$.
  Of course, the result is also a consequence of Krull's principal ideal theorem.)

2.† Let M be a numerical monoid generated by positive weights $\gamma = (\gamma_0, \ldots, \gamma_n)$. In the following we use some concepts and results of Chapter II.

a) The following conditions are equivalent:
(1) M is a complete intersection.
(2)   For every noetherian ring $A \neq 0$ the kernel of $\pi_A^\gamma : A[T] \to A[M]$ is generated by n binomials.
(3)   For every noetherian ring $A \neq 0$ the graded monoid algebra $A[M]$ is a complete intersection over A.
(4)   There is a field K such that the graded monoid algebra $K[M]$ is a complete intersection over K.

b) Assume that A is a noetherian ring $\neq 0$ such that every finite stably free A-module is free (which for instance is true if all finite projective A-modules are free or if the Krull dimension of A is ≤ 1). Equivalent are:
(1) M is a complete intersection.
(5)   For every surjective homogeneous A-algebra homomorphism
$$A[X] = A[X_0, \ldots, X_m] \longrightarrow A[M]$$
the kernel is generated by m homogeneous polynomials. (Suppl. 4 of Sect. 7.)

c) Assume that M is a complete intersection and that for some $A \neq 0$ the kernel of $\pi_A^\gamma : A[T] \to A[M]$ is generated by n homogeneous polynomials of degrees $\delta_1, \ldots, \delta_n$. Then

$$\mathcal{P}_M = \frac{(1-Z^{\delta_1})\cdots(1-Z^{\delta_n})}{(1-Z^{\gamma_0})\cdots(1-Z^{\gamma_n})} \quad \text{and} \quad \mathcal{G}_M = \frac{(1-Z)(1-Z^{\delta_1})\cdots(1-Z^{\delta_n})}{(1-Z^{\gamma_0})(1-Z^{\gamma_1})\cdots(1-Z^{\gamma_n})}.$$

M is symmetric with
$$\mathbf{g}_M = \delta_1 + \cdots + \delta_n - |\gamma|.$$
(See Prop. 7.2 and Suppl. 5 of Sect. 3.) Furthermore, because of $\mathcal{G}_M(1) = 1$
$$\gamma_0 \cdots \gamma_n = \delta_1 \cdots \delta_n.$$

3. Let $M = \text{Mon}(\gamma)$ be a numerical monoid which is a complete intersection. Then
$$\mathcal{R}_\gamma \mathcal{P}_M = \sum_{r=0}^{n-1} \mathcal{P}_{\text{Sat}^r(\gamma)}, \quad \mathcal{R}_\gamma \mathcal{G}_M = \sum_{r=0}^{n-1} \mathcal{G}_{\text{Sat}^r(\gamma)}.$$
(Let $\delta_1, \ldots, \delta_n$ denote the degrees of the minimal relations of $\text{Mon}(\gamma)$ and $I_j := M + \delta_j$. Then $\mathcal{R}_\gamma \mathcal{P}_M = \sum_{j=1}^n \mathcal{P}_{I_j} = \sum_{r=1}^n \mathcal{P}_{\text{Sat}^{n-r}(\gamma)}$ because for any $s \in \mathbb{N}$ the number $c(s)$ of j with $s \in I_j$ coincides by 4.6 with the number $d(s)$, for which $s \in \text{Sat}^{n-d(s)}(\gamma)$, $s \notin \text{Sat}^{n-d(s)-1}(\gamma)$.)

4. Let M be a numerical monoid generated by positive weights $\gamma = (\gamma_0, \ldots, \gamma_n)$.
a) Let $\gamma_{n+1} \in M_+$ and $\bar{\gamma} := (\gamma_0, \ldots, \gamma_n, \gamma_{n+1})$. Then
$$\mathcal{R}_{\bar{\gamma}} = \mathcal{R}_\gamma + Z^{\gamma_{n+1}}.$$
If $\gamma_{n+1} = \langle \alpha, \gamma \rangle$, the new relation is $T_{n+1} - T^\alpha$.

  A simple example is the complete intersection $\mathbb{N}$ generated by $\gamma = (\gamma_0, \ldots, \gamma_n)$ with $\gamma_0 = 1$, where the ideal of relations is generated by $T_j - T_0^{\gamma_j}$, $j = 1, \ldots, n$.

Starting from the uniquely determined system of generators of M the formula $\mathcal{R}_{\bar{\gamma}}(1) = 1 + \mathcal{R}_{\gamma}(1)$ shows in an elementary way that the deviation $\mathrm{dev}_M = \mathrm{dev}_\gamma = \mathcal{R}_\gamma(1)$ is independent of the chosen system γ.

b) Let $\gamma = (\gamma_0, d\gamma_1', \ldots, d\gamma_n')$, $\gamma' := (\gamma_0, \gamma_1', \ldots, \gamma_n')$ be a situation of reduction in codimension 1 as in Section 3. Then
$$\mathcal{R}_\gamma = \mathcal{R}_{\gamma'}(Z^d).$$
(If $d-m$, then $c_m(\gamma) \leq 1$: For every $\alpha \in \mathrm{N}_m(\gamma)$, $\alpha_0 \neq 0$. If $d|m$, $\alpha \in \mathrm{N}_m(\gamma)$ and the corresponding $\alpha' \in \mathrm{N}_{m/d}(\gamma')$ have the same support. Thus $c_m(\gamma) = c_{m/d}(\gamma')$.)

c) Compute $\mathcal{R}_\gamma$ for $\gamma = (24, 32, 40, 26, 39)$ using a) and b).

5. Let M be a numerical monoid generated by positive weights $\gamma = (\gamma_0, \ldots, \gamma_n)$.

a) $g_\gamma + \max_i \gamma_i \ \leq \ \mathrm{con}_\gamma \ \leq \ g_\gamma + \min_i \gamma_i + \max_i \gamma_i$.

$(g_\gamma + \gamma_i \notin M + \gamma_i \supseteq \mathrm{Con}(\gamma)$.) If $g_\gamma + \max_i \gamma_i < \mathrm{con}_\gamma$, then $\mathrm{rmax}_\gamma = \mathrm{con}_\gamma$. (The equality $g_\gamma + \max_i \gamma_i = \mathrm{rmax}_\gamma = \mathrm{con}_\gamma$ may happen, e. g. for $\gamma = (3, 4, 5, 8)$ or $\gamma = (7, 12, 15, 20)$.)

b) Let E_i be the least positive multiple of γ_i contained in $\mathrm{Mon}(\gamma_j : j \neq i)$ and $E_i = v_i \gamma_i = \sum_{j \neq i} \alpha_{ij} \gamma_j$ with $v_i, \alpha_{ij} \in \mathbb{N}$. Then the binomial
$$T_i^{v_i} - \prod_{j \neq i} T_j^{\alpha_{ij}}$$
belongs to a minimal set of binomial generators of the Ideal $\mathfrak{I}$ of relations of M with respect to γ. In particular, $E_i \notin \mathrm{Con}(\gamma)$ and
$$\max_i \gamma_i \ \leq \ \max_i E_i \ \leq \ \mathrm{rmax}_\gamma \ \leq \ \mathrm{con}_\gamma.$$
($\{i\}$ is a connected component of $\mathrm{G}_{E_i}(\gamma) \neq \{i\}$.) The E_i are called the c o r n e r s and the specified binomials of degree E_i the c o r n e r   b i n o m i a l s of γ.

c) Let $\bar{\gamma} = (\gamma_0, .., \gamma_n, \gamma_{n+1})$ with $\gamma_{n+1} > 0$. Then
$$\mathrm{Con}(\bar{\gamma}) \ \supseteq \ \mathrm{Con}(\gamma) \cap (M_+ + \gamma_{n+1});$$
equality holds if $\gamma_{n+1} \in M$. In particular,
$$\mathrm{con}_{\bar{\gamma}} \ \leq \ \max(\mathrm{con}_\gamma, g_{M_+} + \gamma_{n+1});$$
equality holds if $\gamma_{n+1} \in M$. (Let $\gamma_{n+1} = \beta_0 \gamma_0 + \cdots + \beta_n \gamma_n$. If $\alpha \in \mathrm{N}_m(\bar{\gamma})$ then $(\alpha_0, \ldots, \alpha_n) + \alpha_{n+1}(\beta_0, \ldots, \beta_n) \in \mathrm{N}_m(\gamma)$.)

d) Suppose $\gamma_0 \leq \gamma_i \leq \gamma_n$ for all i and $\mathrm{con}_\gamma > g_\gamma + \gamma_n$. Then $\mathrm{con}_\gamma = \mathrm{rmax}_\gamma$. Furthermore, $\mathrm{con}_\gamma = -t + \gamma_0 + \gamma_k$ where t is an element of the type set $\mathrm{T}(M)$ and $\gamma_n + g_\gamma - (-t) < \gamma_0 + \gamma_k$, $\gamma_0 < \gamma_k$. (For γ_k one can take any element for which k does not belong to the connected component of 0 in $\mathrm{G}_m(\gamma)$, $m := \mathrm{con}_\gamma$. For an example consider $\gamma = (6, 8, 9, 11, 13)$ where $\mathrm{con}_\gamma = 26$, $g_\gamma = 10$, $-t = 7$.)

6. Let M be a numerical monoid generated by positive weights $\gamma = (\gamma_0, \ldots, \gamma_n)$. By $\delta_1, \ldots, \delta_r$ we denote the degrees of the elements of a minimal system $F_1, \ldots, F_r$ of homogeneous generators of $\mathfrak{I} = \ker \pi_K^\gamma$, K a field and $\pi_K^\gamma : K[T] \to K[M]$ the canonical homomorphism. Then rmax_γ is the supremum of $\delta_1, \ldots, \delta_r$.

a) An element $m \in \mathbb{N}$ belongs to $\mathrm{VSat}(\gamma)$ if and only if $\mathfrak{I}_m K[T^{\pm 1}] = \mathfrak{I}K[T^{\pm 1}]$. (4.4(1).) From this follows
$$\mathrm{VSat}(\gamma) \ \supseteq \ \bigcap_{j=1}^r (M + \delta_j) \quad \text{and} \quad \mathrm{vsat}_\gamma \ \leq \ g_\gamma + \mathrm{rmax}_\gamma.$$

b) If M is a complete intersection (i. e. $r = n$), then
$$\mathrm{Sat}(\gamma) \ = \ \mathrm{VSat}(\gamma) \ = \ \bigcap_{j=1}^n (M + \delta_j) \quad \text{and} \quad \mathrm{sat}_\gamma \ = \ \mathrm{vsat}_\gamma \ = \ g_\gamma + \mathrm{rmax}_\gamma.$$
(This follows partly from 4.6, too. The equalities $\mathrm{sat}_\gamma = \mathrm{vsat}_\gamma = g_\gamma + \mathrm{rmax}_\gamma$ may hold if M is not a complete intersection, e. g. $\gamma = (6, 7, 8, 9)$, but in this example

$\mathrm{Sat}(\gamma) = \mathrm{VSat}(\gamma) \neq \bigcap_{j=1}^{r}(M + \delta_j)$, $r = 4$, $\delta = (14, 15, 16, 18)$. See also Suppl. 1 in Sect. 5 and its information about rmax_γ, con_γ, sat_γ and vsat_γ.)

7. Let M be a numerical monoid generated by positive weights $\gamma = (\gamma_0, \ldots, \gamma_n)$, $n \geq 1$. Assume that there is an element $b \in \mathbb{N}$ such that $G_b(\gamma) = \{0, \ldots, n\}$ and that every component of $G_b(\gamma)$ is a single point. Then $b = a_i\gamma_i$, $i = 0, \ldots, n$, with $a_i \in \mathbb{N}^*$. The elements $a_0, \ldots, a_n$ are pairwise relatively prime, $b = a_0 \cdots a_n$ and the ideal $\mathfrak{I}$ of relations of M with respect to γ is (minimally) generated by the binomials $F_j = T_0^{a_0} - T_j^{a_j}$, $j = 1, \ldots, n$. In particular, M is a complete intersection. Its Poincaré series (cf. Suppl. 2) is
$$\mathcal{P}_M = (1 - Z^b)^n / (1 - Z^{b/a_0}) \cdots (1 - Z^{b/a_n}).$$
(For $i \neq j$ one has $b = k \cdot \mathrm{lcm}(\gamma_i, \gamma_j)$ and necessarily $k = 1$, since otherwise the elements i, j would be connected. Then obviously $\gcd(a_i, a_j) = 1$ and $b = a_0 \cdots a_n$. Thus, M is a monoid of the type treated in Sect. 3, Suppl. 12b). To prove that $\mathfrak{I}$ is generated by $F_1, \ldots, F_n$ it is enough now to show that the M-ideal $\mathrm{Sat}^{n-1}(\gamma)$ is generated by b. Consider $m \in \mathrm{Sat}^{n-1}(\gamma)$. There are two different representations $m = \langle \alpha, \gamma \rangle = \langle \beta, \gamma \rangle$ with $\alpha, \beta \in \mathbb{N}^{n+1}$. There is some index i with $\alpha_i > \beta_i$, say $i = 0$. Then $(\alpha_0 - \beta_0)\gamma_0 \in \mathbb{Z}\gamma_1 + \cdots + \mathbb{Z}\gamma_n = \mathbb{Z}a_0$, and there is an element $c \in \mathbb{N}^*$ such that $\alpha_0 = \beta_0 + ca_0$. Thus
$$m = \beta_0\gamma_0 + ca_0\gamma_0 + \alpha_1\gamma_1 + \cdots + \alpha_n\gamma_n \in cb + M \subseteq b + M.$$
Conversely, if $\mathrm{Sat}^{n-1}(\gamma)$ is a principal $\mathrm{Mon}(\gamma)$-ideal, $\gamma = (\gamma_0, \ldots, \gamma_n)$ arbitrary, $n \geq 1$, with generator b', then the ideal $\mathfrak{I}$ of relations is generated by binomials of degree b'. Therefore, we are necessarily in the situation discussed above with $b = b'$. This was also proved in [17], Beispiel 5.12. Another characterization of the situation above is to say that all the relations are generated by relations of the same degree.)

8.† Let M be a numerical monoid generated by positive weights $\gamma = (\gamma_0, \ldots, \gamma_n)$, $n \geq 2$. Assume that there is an element $b \in \mathbb{N}$ such $G_b(\gamma) = \{1, \ldots, n\}$ and that every component of $G_b(\gamma)$ is a single point. Let $d := \gcd(\gamma_1, \ldots, \gamma_n)$. Then $b = a_i\gamma_i$, $i = 1, \ldots, n$, with $a_i \in \mathbb{N}^*$. The elements $a_1, \ldots, a_n$ are pairwise relatively prime and $b = da_1 \cdots a_n$. Furthermore, let E_0 be the least positive multiple of γ_0 contained in $\mathrm{Mon}(\gamma_1, \ldots, \gamma_n)$, $E_0 = v_0\gamma_0 = \alpha_1\gamma_1 + \cdots + \alpha_n\gamma_n$. Then $E_0 \neq b$ and the binomials
$$T_1^{a_1} - T_j^{a_j}, \quad 2 \leq j \leq n, \qquad T_0^{v_0} - T_1^{\alpha_1} \cdots T_n^{\alpha_n}$$
are part of a minimal system of binomial generators of $\mathfrak{I}$. These binomials form a full system of generators of $\mathfrak{I}$, i.e. M is a complete intersection, if and only if $v_0 = d$. In case $n = 2$ the last condition is always fulfilled. (We give only a hint for the case $n = 2$. Let $\gamma'_i := \gamma_i/d$, $i = 1, 2$, and assume $\gamma_0 \notin \mathrm{Mon}(\gamma'_1, \gamma'_2)$. Then there are $a_1, a_2 \in \mathbb{N}$ such that
$$\gamma_0 + a_1\gamma'_1 + a_2\gamma'_2 = g_{(\gamma'_1, \gamma'_2)} = \gamma'_1\gamma'_2 - \gamma'_1 - \gamma'_2.$$
Thus $b = d\gamma'_1\gamma'_2 = d\gamma_0 + (a_1 + 1)\gamma_1 + (a_2 + 1)\gamma_2$, which contradicts the structure of $G_b(\gamma)$. Remark: If $n \geq 3$, in general M is not a complete intersection, e. g. for $\gamma = (11, 6, 10, 15)$.)

9.† In this supplement we consider numerical monoids generated by three positive weights $\gamma = (\gamma_0, \gamma_1, \gamma_2)$. The structure of such monoids can be described explicitly, which was done by S.M. Johnson [20], J. Herzog [16] and J. Kraft [29]. The following approach includes their results. The ways to carry out actual computations have been studied for a long time, too, starting with observations by Johnson; for this topic see J.L. Davison [10]. Let the ground ring A be a field K.

There are three partition types for a set $G_m := G_m(\gamma)$ with $c_m(\gamma) > 1$, $m \in \mathbb{N}$, which can be indicated in the following self-explanatory way: $\odot \odot \odot$, $\cdot \odot \odot$, $\odot \odot\!\!\!\!\odot$. In the cases where there is an element $m \in \mathbb{N}$ such that G_m is of type $\odot \odot \odot$ or of type $\cdot \odot \odot$, the monoid M is a complete intersection by Suppls. 7 and 8. Therefore, *we will assume from now on that the non-empty, non-connected G_m are of type $\odot \odot\!\!\!\!\odot$ only.*

a) At the corners E_0, E_1, E_2 each G_{E_i} is non-connected. In our special situation therefore these corners are pairwise different and in particular M is not a complete intersection. Fix representations $E_0 = v_0\gamma_0 = a_1\gamma_1 + a_2\gamma_2$, $E_1 = v_1\gamma_1 = b_0\gamma_0 + b_2\gamma_2$, $E_2 = v_2\gamma_2 = c_0\gamma_0 + c_1\gamma_1$ and let $\mathfrak{S}$ be the matrix

$$\mathfrak{S} := \begin{pmatrix} v_0 & -b_0 & -c_0 \\ -a_1 & v_1 & -c_1 \\ -a_2 & -b_2 & v_2 \end{pmatrix}.$$

The corresponding corner binomials

$$F_0 := T_0^{v_0} - T_1^{a_1}T_2^{a_2}, \quad F_1 := T_1^{v_1} - T_0^{b_0}T_2^{b_2}, \quad F_2 := T_2^{v_2} - T_0^{c_0}T_1^{c_1}$$

form a minimal system of generators of $\mathfrak{I}$. (Otherwise there would be an element $m \notin \{E_0, E_1, E_2\}$ such that G_m is of type $\odot \odot\!\!\!\!\odot$. We may assume $m = u\gamma_0 = v\gamma_1 + w\gamma_2$. Then $u > v_0$ and $m = (u - v_0)\gamma_0 + (v + a_1)\gamma_1 + (w + a_2)\gamma_2$, a contradiction.) In particular

$$\mathcal{R}_\gamma = Z^{E_0} + Z^{E_1} + Z^{E_2}.$$

(In general, for any almost arithmetic sequence γ of weights an explicit minimal set of binomial relations can be constructed, cf. Patil [32]. $\gamma = (\gamma_0, \ldots, \gamma_n)$ is called an **a l m o s t   a r i t h m e t i c   s e q u e n c e** if n of the weights form an arithmetic sequence.)

b) First we develop some properties of the matrix $\mathfrak{S}$, also proved by S.M. Johnson and J. Herzog: All the entries of $\mathfrak{S}$ are non-zero. The sum of its columns is the zero-column. $\mathfrak{S}$ is uniquely determined, i.e. the corner binomials are uniquely determined.

(Assume $a_2 = 0$, i.e. $E_0 = v_0\gamma_0 = a_1\gamma_1$. Then $a_1 > v_1$ because $E_0 \neq E_1$. There is another representation $E_0 = a_1'\gamma_1 + a_2'\gamma_2$ with $a_2' \neq 0$ because of the type of G_{E_0}. Similarly, there is a representation $E_1 = b_0'\gamma_0 + b_2'\gamma_2$ with $b_0' \neq 0$. Now $(a_1 - a_1')\gamma_1 = a_2'\gamma_2$ and therefore $a_1 - a_1' \geq v_1$, say $a_1 = v_1 + a_1' + r$. It follows that $(v_0 - b_0')\gamma_0 = (a_1' + r)\gamma_1 + b_2'\gamma_2 > 0$ and $v_0 - b_0' < v_0$, contradiction.

We have $0 < (b_0 + c_0)\gamma_0 = (v_1\gamma_1 - b_2\gamma_2) + (v_2\gamma_2 - c_1\gamma_1) = (v_1 - c_1)\gamma_1 + (v_2 - b_2)\gamma_2$. If $v_1 - c_1 < 0$ then $v_2 - b_2 > 0$, $(v_2 - b_2)\gamma_2 = (b_0 + c_0)\gamma_0 + (c_1 - v_1)\gamma_1 \in \mathrm{Mon}(\gamma_0, \gamma_1)$ and $v_2 > v_2 - b_2$, contradiction. Hence $v_1 - c_1 \geq 0$, $v_2 - b_2 \geq 0$ and $b_0 + c_0 \geq v_0$. Similarly $a_1 + c_1 \geq v_1$ and $a_2 + b_2 \geq v_2$. This means that all the elements of the sum of the columns of $\mathfrak{S}$ are non-positive. Since this sum is a syzygy of $(\gamma_0, \gamma_1, \gamma_2)$, it is zero.

Because the sum of the columns of $\mathfrak{S}$ is zero for any choice of the elements $a_1, a_2, b_0, b_2, c_0, c_1$, these elements are unique.)

c) In $K[T]$ the corner binomials are (non-associated) prime polynomials. (Every element of a minimal generating system of homogeneous polynomials of the prime ideal $\mathfrak{I}$ is prime.)

The Lasker-Noether decomposition of the ideal (F_0, F_1) in $K[T]$ is

$$(F_0, F_1) = \Im \cap (T_0^{b_0}, T_1^{a_1}),$$

where the ideal $(T_0^{b_0}, T_1^{a_1})$ is primary to (T_0, T_1), and similarly for (F_0, F_2), (F_1, F_2). (Obviously, (F_0, F_1) is contained in the intersection. For the converse, it remains to prove: If $H \cdot F_2 \in (T_0^{b_0}, T_1^{a_1})$ then $H \in (F_0, F_1)$. But obviously $T_0^{b_0}, T_1^{a_1}, F_2$ is a regular sequence. Therefore $H \in (T_0^{b_0}, T_1^{a_1})$. Thus we have to show that $(T_0^{b_0}, T_1^{a_1})F_2 \subseteq (F_0, F_1)$. But, using b) one gets simply:

$$T_1^{c_1} F_0 + T_2^{a_2} F_1 + T_0^{b_0} F_2 = 0, \quad T_2^{b_2} F_0 + T_0^{c_0} F_1 + T_1^{a_1} F_2 = 0 .)$$

d) As an application of the primary decomposition in c) we get the following identity in $K[T^{\pm 1}]$:

$$(F_0, F_1) \cdot K[T^{\pm 1}] = \Im K[T^{\pm 1}].$$

By Theorem 4.4(1) this is equivalent with the fact that the syzygies $(v_0, -a_1, -a_2)$ and $(-b_0, v_1, -b_2)$ form a basis of $\mathrm{Syz}(\gamma)$, i.e. that the 2×2-minors of the first two columns of the matrix $\mathfrak{S}$ are relatively prime. The analogue holds for the minors of any pair of columns of $\mathfrak{S}$. From that follows for the adjoint of $\mathfrak{S}$:

$$\mathrm{adj}\,\mathfrak{S} = \begin{pmatrix} \gamma_0 & \gamma_1 & \gamma_2 \\ \gamma_0 & \gamma_1 & \gamma_2 \\ \gamma_0 & \gamma_1 & \gamma_2 \end{pmatrix}.$$

In particular

$$\begin{vmatrix} -a_1 & v_1 \\ -a_2 & -b_2 \end{vmatrix} = \gamma_0, \quad -\begin{vmatrix} v_0 & -b_0 \\ -a_2 & -b_2 \end{vmatrix} = \gamma_1, \quad \begin{vmatrix} v_0 & -b_0 \\ -a_1 & v_1 \end{vmatrix} = \gamma_2.$$

(Because $(\mathrm{adj}\,\mathfrak{S}) \cdot \mathfrak{S} = 0$, every row of $\mathrm{adj}\,\mathfrak{S}$ is a syzygy of the columns of $\mathfrak{S}$. Since the components of such a row are relatively prime, every row coincides up to sign with $(\gamma_0, \gamma_1, \gamma_2)$. From $\mathfrak{S} \cdot (\mathrm{adj}\,\mathfrak{S}) = 0$ and $\mathfrak{S} \cdot {}^t(1, 1, 1) = 0$ follows that every column of $\mathrm{adj}\,\mathfrak{S}$ is constant. So it remains to show that one element of $\mathrm{adj}\,\mathfrak{S}$ is non-negative. But, e.g. $(-a_1)(-b_2) - v_1(-a_2) = a_1 b_2 + v_1 a_2 \geq 0$.)

e) The ideals

$$\Im \cap (T_0^{b_0}, T_1^{a_1}) = (F_0, F_1), \quad \Im + (T_0^{b_0}, T_1^{a_1}) = (T_0^{b_0}, T_1^{a_1}, F_2), \quad (T_0^{b_0}, T_1^{a_1})$$

are generated by regular sequences. The canonical Mayer-Vietoris sequence

$$0 \longrightarrow K[T]/\Im \cap (T_0^{b_0}, T_1^{a_1}) \longrightarrow (K[T]/\Im) \oplus (K[T]/(T_0^{b_0}, T_1^{a_1}))$$
$$\longrightarrow K[T]/(\Im + (T_0^{b_0}, T_1^{a_1})) \longrightarrow 0$$

yields for $\mathcal{P}_M = \mathcal{P}_{K[T]/\Im}$ the identity

$$\mathcal{P}_M = \frac{P}{(1 - Z^{\gamma_0})(1 - Z^{\gamma_1})(1 - Z^{\gamma_2})} = \frac{1 - \mathcal{R}_\gamma + Z^{D_1} + Z^{D_2}}{(1 - Z^{\gamma_0})(1 - Z^{\gamma_1})(1 - Z^{\gamma_2})}$$

with

$$P = (1 - Z^{E_0})(1 - Z^{E_1}) + (1 - Z^{b_0 \gamma_0})(1 - Z^{a_1 \gamma_1})(1 - Z^{E_2})$$
$$- (1 - Z^{b_0 \gamma_0})(1 - Z^{a_1 \gamma_1})$$
$$= (1 - Z^{E_0})(1 - Z^{E_1}) - (1 - Z^{b_0 \gamma_0})(1 - Z^{a_1 \gamma_1})Z^{E_2}$$
$$= 1 - (Z^{E_0} + Z^{E_1} + Z^{E_2}) + Z^{D_1} + Z^{D_2}$$

as numerator, where the exponents
$$D_1 := b_0\gamma_0 + E_2, \quad D_2 := a_1\gamma_1 + E_2$$
are greater than E_0, E_1, E_2. It follows that the Frobenius number of M is
$$\mathrm{g}_M = \deg \mathcal{P}_M = \max(D_1, D_2) - |\gamma|.$$
Furthermore, $D_1 \neq D_2$, and D_1, D_2 are related by
$$D_1 + D_2 = E_0 + E_1 + E_2, \quad D_1 \cdot D_2 = E_0 E_1 + E_0 E_2 + E_1 E_2 - \gamma_0\gamma_1\gamma_2.$$
($D_1 = a_2\gamma_2 + E_1 = c_1\gamma_1 + E_0$ and $D_2 = c_0\gamma_0 + E_1 = b_2\gamma_2 + E_0$ by b). Useful are
the relations $b_0\gamma_0 + a_1\gamma_1 = E_0 + E_1 - E_2$ and $b_0\gamma_0 a_1\gamma_1 = E_0 E_1 - \gamma_0\gamma_1\gamma_2$. Note
$b_0 a_1 = v_0 v_1 - \gamma_2$, cf. d).)

f) The degree of singularity of M is determined by the following formula of
J. Kraft:
$$2\vartheta_M - 1 = E_0 + E_1 + E_2 - |\gamma| - \frac{E_0 E_1 E_2}{\gamma_0\gamma_1\gamma_2}.$$
($\vartheta_M = \mathcal{G}'_M(1) = \mathcal{G}'_M(1)/\mathcal{G}_M(1)$ by Sect. 3. One employs the formula
$$\frac{\mathcal{H}'(1)}{\mathcal{H}(1)} = \frac{1}{2}(p + \deg \mathcal{H}),$$
which holds for any rational function of type
$$\mathcal{H} = Z^p \frac{(1 - Z^{m_1}) \cdots (1 - Z^{m_r})}{(1 - Z^{n_1}) \cdots (1 - Z^{n_r})}$$
with $m_1, \ldots, m_r, n_1, \ldots, n_r \in \mathbb{Z} \smallsetminus \{0\}$, $p \in \mathbb{Z}$, and uses the hints given in e).)

g) The type set of M, i.e. the set of minimal generators of Ω_M, is
$$\mathrm{T}(M) = \{-D_1 + |\gamma|, -D_2 + |\gamma|\}.$$
(One way to prove this is by using simple direct computations similar to those
used in c). We prefer to give a different proof which sheds more light on the
nature of the whole set-up.

Let Ω denote the M-ideal $(-D_1 + |\gamma| + M) \cup (-D_2 + |\gamma| + M)$. We will prove
that $\mathcal{P}_\Omega = -\mathcal{P}_M(1/Z) = \mathcal{P}_{\Omega_M}$, from which $\Omega = \Omega_M$ follows.

$\mathcal{P}_\Omega$ is equal to the Poincaré function $\mathcal{P}_\omega$ of the graded $K[M]$-module $\omega :=$
$K[M]x_1 + K[M]x_2 \subseteq K[X^{\pm 1}]$, where
$$x_1 := X^{-D_1 + |\gamma|}, \quad x_2 := -X^{-D_2 + |\gamma|}.$$
The syzygies of x_1, x_2 are generated by the columns of the matrix
$$\mathfrak{A} := \begin{pmatrix} T_1^{c_1} & T_2^{a_2} & T_0^{b_0} \\ T_2^{b_2} & T_0^{c_0} & T_1^{a_1} \end{pmatrix}.$$
That the columns are syzygies of x_1, x_2, can be easily checked. Consider any
relation $H_1 x_1 + H_2 x_2 = 0$, $H_1, H_2 \in K[T]$. Multiplying by $X^{D_1 + D_2 - E_2 - |\gamma|}$ yields
$H_1 T_1^{a_1} - H_2 T_0^{b_0} \in \mathfrak{I} \cap (T_0^{b_0}, T_1^{a_1}) = (F_0, F_1)$, from which $H_1 \in (T_1^{c_1}, T_2^{a_2}, T_0^{b_0})$
follows. Thus the columns of $\mathfrak{A}$ generate the syzygies of x_1, x_2 over $K[T]$.

The syzygies of the columns of $\mathfrak{A}$ are simply $K[T] \cdot (F_0, F_1, F_2)$: Apply Cramer's
rule. Thus there is a natural exact sequence
$$0 \longrightarrow K[T](-|\gamma|) \longrightarrow \bigoplus_{i=0}^{2} K[T](E_i - |\gamma|) \longrightarrow \bigoplus_{i=1}^{2} K[T](D_i - |\gamma|) \longrightarrow \omega \longrightarrow 0$$
of graded $K[T]$-modules and
$$\mathcal{P}_\omega = \mathcal{P}_{K[T]} \cdot Z^{|\gamma|}(Z^{-D_1} + Z^{-D_2} - (Z^{-E_0} + Z^{-E_1} + Z^{-E_2}) + 1),$$

hence $\mathcal{P}_\omega = -\mathcal{P}_M(1/Z)$, which finishes the proof.

The syzygies of F_0, F_1, F_2 are generated by the rows of $\mathfrak{A}$, which are linearly independent. There is a corresponding natural exact sequence

$$0 \longrightarrow \bigoplus_{i=1}^{2} K[T](-D_i) \longrightarrow \bigoplus_{i=0}^{2} K[T](-E_i) \longrightarrow K[T] \longrightarrow K[M] \longrightarrow 0$$

of graded $K[T]$-modules, which yields the (already known) formula for $\mathcal{P}_M$. In addition, this sequence can be used to compute the dualizing module $\omega_{K[M]}$, which is isomorphic to $\operatorname{Ext}^2_{K[T]}(K[M], K[T](-|\gamma|))$ and hence isomorphic to the residue class module of $K[T](D_1 - |\gamma|) \oplus K[T](D_2 - |\gamma|)$ modulo the columns of $\mathfrak{A}$. Thus $\omega \cong \omega_{K[M]}$ is indeed a model for the dualizing module of the Macaulay ring $K[M]$, with corresponding M-ideal Ω_M.)

(*Remark.* The formulas for g_M and ϑ_M hold also in the case of a complete intersection generated by $\gamma_0, \gamma_1, \gamma_2$. This is easily verified using the explicit descriptions in Suppls. 7,8. Note that also in this case the numerator of $\mathcal{P}_M$ can be written in the form $1 - \mathcal{R}_\gamma + Z^{D_1} + Z^{D_2}$ with well-defined different positive integers D_1, D_2. Assume $D_1 < D_2$. Then $D_2 = g_M + |\gamma| = E_0 + E_1$ and $\gamma_0 \gamma_1 \gamma_2 = E_0 E_1$ if $E_0 = E_2$. The type set of M is simply $\{-D_2 + |\gamma|\}$. The element $-D_1 + |\gamma|$ belongs here to $\Omega_M \setminus \mathrm{T}(M)$.)

5 Splitting of Numerical Monoids

A useful tool in the handling of numerical monoids is a *splitting procedure*, first considered by K. Watanabe and Ch. Delorme, cf. [11].

As in the last sections we consider a numerical monoid $M = \text{Mon}(\gamma)$ generated by positive weights $\gamma = (\gamma_0, \ldots, \gamma_n)$. Let $m \in \mathbb{N}$ and let us assume that $\Omega := \{0, \ldots, n\}$ is decomposed into two disjoint sets: $\Omega = \Omega' \uplus \Omega''$ such that both $\Omega' \cap G_m(\gamma)$ and $\Omega'' \cap G_m(\gamma)$ are non-empty unions of connected components of $G_m(\gamma)$; for the definition of $G_m(\gamma)$ see the remarks following Proposition 4.3.

Furthermore, define $a' := \gcd(\gamma_i : i \in \Omega')$, $\gamma' := (\gamma_i/a')_{i \in \Omega'}$, $M' := \text{Mon}(\gamma')$, $\Gamma' := \text{Syz}(\gamma')$ [1] and similarly a'', γ'', M'', Γ'' with respect to Ω''. Then $M = a'M' + a''M''$, $m = a'm' = a''m''$ with $m' \in M'$, $m'' \in M''$ and $m = a'a''d$ with $d = \gcd(m', m'')$.

5.1 Proposition *In the situation just described, there is a canonical exact sequence of abelian groups*

$$0 \longrightarrow (\Gamma'/\Gamma(m';\gamma')) \oplus (\Gamma''/\Gamma(m'';\gamma'')) \longrightarrow \Gamma/\Gamma(m;\gamma) \longrightarrow \mathbb{Z}/\mathbb{Z}d \longrightarrow 0.$$

In particular, $m \in \text{Sat}(\gamma)$ if and only if $m' \in \text{Sat}(\gamma')$ and $m'' \in \text{Sat}(\gamma'')$. In that case

$$[\Gamma : \Gamma(m;\gamma)] = d \cdot [\Gamma' : \Gamma(m';\gamma')] \cdot [\Gamma'' : \Gamma(m'';\gamma'')].$$

Furthermore, $m \in \text{VSat}(\gamma)$ if and only if $d = 1$ and $m' \in \text{VSat}(\gamma')$, $m'' \in \text{VSat}(\gamma'')$.

Proof. We will use the canonical decomposition $\mathbb{Z}^\Omega = \mathbb{Z}^{\Omega'} \oplus \mathbb{Z}^{\Omega''}$. Choosing elements $\alpha' \in \mathbb{N}_{m'}(\gamma')$, $\alpha'' \in \mathbb{N}_{m''}(\gamma'')$ one has

$$\Gamma(m;\gamma) = \mathbb{Z}\alpha + \Gamma(m';\gamma') + \Gamma(m'';\gamma''),$$

where $\alpha := (\alpha', -\alpha'') = \alpha' - \alpha'' \in \mathbb{Z}^\Omega = \mathbb{Z}^{\Omega'} \oplus \mathbb{Z}^{\Omega''}$. The (surjective) linear form $\gamma : \mathbb{Z}^\Omega \to \mathbb{Z}$ is the composition

$$\mathbb{Z}^{\Omega'} \oplus \mathbb{Z}^{\Omega''} \xrightarrow{\gamma' \oplus \gamma''} \mathbb{Z} \oplus \mathbb{Z} \xrightarrow{(a', a'')} \mathbb{Z}.$$

From the canonical commutative diagram

$$
\begin{array}{ccccccccc}
0 & \to & \Gamma(m';\gamma') \oplus \Gamma(m'';\gamma'') & \longrightarrow & \Gamma(m;\gamma) & \longrightarrow & \mathbb{Z}\alpha & \to & 0 \\
 & & \downarrow & & \downarrow & & \downarrow & & \\
0 & \to & \Gamma' \oplus \Gamma'' & \longrightarrow & \Gamma & \longrightarrow & \mathbb{Z}(a'', -a') & \to & 0
\end{array}
$$

---

[1] Recall that we sometimes use the abbreviation $\Gamma = \text{Syz}(\gamma)$.

with exact rows one gets the desired exact sequence of the Proposition as the exact sequence of the cokernels, because the image of α in $\mathbb{Z}(a'', -a')$ is the element $(m', -m'') = d(a'', -a')$. □

Let us say that $M = \text{Mon}(\gamma)$ **splits along** Ω', Ω'' if

$$M = a'M' + a''M'',$$

where $a' \in M'' = \text{Mon}(\gamma'')$ and $a'' \in M' = \text{Mon}(\gamma')$. Then we are in the situation studied above for $m = a'a''$, $d = 1$, $m' = a''$, $m'' = a'$, because the support of any $\alpha \in N_{a'a''}(\gamma)$ necessarily is contained in Ω' or in Ω''. In this case *the monoid M is canonically isomorphic to the quotient $(M' \oplus M'')/R$, where R is the compatible equivalence relation generated by the single relation $((m',0),(0,m'')) = ((a'',0),(0,a'))$.* Proof. Under the surjective homomorphism

$$M' \oplus M'' \xrightarrow{(a',a'')} M$$

by $(x,y) \mapsto a'x + a''y$ two elements (x_1, y_1), (x_2, y_2) have the same image if and only if $(x_1, y_1) = (x_2, y_1) + k(m', 0)$, $(x_2, y_2) = (x_2, y_1) + k(0, m'')$ or $(x_1, y_1) = (x_1, y_2) + k(0, m'')$, $(x_2, y_2) = (x_1, y_2) + k(m', 0)$ for some $k \in \mathbb{N}$. To put it in a different way: M is isomorphic to $(a'M' \oplus a''M'')/\tilde{R}$, where $\tilde{R}$ is generated by $((m,0),(0,m))$, $m = a'a''$.

This description of M shows that there is a canonical isomorphism

$$K[a'M'] \otimes_{K[X^m]} K[a''M''] =$$
$$K[a'M'] \otimes_K K[a''M'']/(X^m \otimes 1 - 1 \otimes X^m) = K[M]$$

of graded K-algebras. It follows:

5.2 Theorem *If M splits in the way described above, then*

$$\mathcal{P}_M = (1 - Z^{a'a''})\mathcal{P}_{M'}(Z^{a'})\mathcal{P}_{M''}(Z^{a''}),$$

$$g_M = a'g_{M'} + a''g_{M''} + a'a'', \qquad \vartheta_M = a'\vartheta_{M'} + a''\vartheta_{M''} + \frac{(a'-1)(a''-1)}{2}.$$

Proof. The formula for $\mathcal{P}_M$ follows from the facts, that $\mathcal{P}_{M'}(Z^{a'})\mathcal{P}_{M''}(Z^{a''})$ is the Poincaré series of the tensor product and that $X^m \otimes 1 - 1 \otimes X^m$ is a non-zero-divisor (because the tensor product is an integral domain contained in $K[a'\mathbb{N}] \otimes_K K[a''\mathbb{N}]$). Multiplying by $1 - Z$ yields

$$\mathcal{G}_M = \frac{(1-Z)(1-Z^{a'a''})}{(1-Z^{a'})(1-Z^{a''})}\mathcal{G}_{M'}(Z^{a'})\mathcal{G}_{M''}(Z^{a''}).$$

A comparison of degrees proves the formula for g_M. The formula for ϑ_M follows easily from $\vartheta_M = \mathcal{G}'_M(1) = \mathcal{G}'_M(1)/\mathcal{G}_M(1)$, using logarithmic derivatives. □

The formulas for g_M and ϑ_M in Theorem 5.2 imply

$$2\vartheta_M - 1 - g_M \;=\; a'(2\vartheta_{M'} - 1 - g_{M'}) + a''(2\vartheta_{M''} - 1 - g_{M''})\,.$$

From this follows directly, that the splitting monoid M is symmetric if and only if its components M' and M'' are both symmetric. This application as well as the formula for g_M were already proved in [11], Proposition 10 in a direct way. More generally, the representation of $K[M]$ from which Theorem 5.2 was derived allows to prove that the type of M is the product of the types of M' and M'', i.e. $t_M = t_{M'}t_{M''}$. This can also be proved directly in a straightforward way.

Coming back to the aforementioned representation of $K[M]$ we note first, that the kernel $\tilde{\mathfrak{J}}'$ of $K[T_i \,:\, i \in \Omega'] \to K[a'M']$ is the kernel $\mathfrak{J}'$ of $K[T'] \to K[M']$, subjected to the isomorphism $K[T'] \cong K[T_i \,:\, i \in \Omega']$ by $T_i' \mapsto T_i$ (which multiplies degrees by a'). Analogously, the kernel $\tilde{\mathfrak{J}}''$ of $K[T_i \,:\, i \in \Omega''] \to K[a''M'']$ can be identified with the kernel $\mathfrak{J}''$ of $K[T''] \to K[M'']$. Thus the kernel $\mathfrak{J}$ of $K[T] \to K[M]$ is generated by $\tilde{\mathfrak{J}}'$, $\tilde{\mathfrak{J}}''$ and the element $F := T^{\alpha'} \otimes 1 - 1 \otimes T^{\alpha''}$, where $\alpha' \in N_{m'}(\gamma') = N_m(a'\gamma)$ and $\alpha'' \in N_{m''}(\gamma'') = N_m(a''\gamma)$ are fixed. Obviously, minimal systems of binomial generators of $\tilde{\mathfrak{J}}'$ and $\tilde{\mathfrak{J}}''$ together with F make up a minimal system of binomial generators of $\mathfrak{J}$. In particular:

5.3 Proposition *In the situation just described,*

$$\mathcal{R}_\gamma \;=\; \mathcal{R}_{\gamma'}(Z^{a'}) + \mathcal{R}_{\gamma''}(Z^{a''}) + Z^{a'a''}\,, \qquad \mathrm{dev}_\gamma \;=\; \mathrm{dev}_{\gamma'} + \mathrm{dev}_{\gamma''}\,.$$

Recall that $\mathrm{dev}_\gamma = \mathcal{R}_\gamma(1) - n$. As a corollary one gets that the splitting monoid M is a complete intersection if and only if both its parts M' and M'' are complete intersections. This was also proved in [11], using slightly different methods.

Supplements

1. Let M be a numerical monoid generated by positive weights $\gamma = (\gamma_0, \ldots, \gamma_n)$.
a) If $\mathrm{rmax}_\gamma \in \mathrm{VSat}(\gamma)$, then $\mathrm{sat}_\gamma = \mathrm{vsat}_\gamma = g_\gamma + \mathrm{rmax}_\gamma$.
(One is in the situation of 5.1 with $d = 1$ and $\mathrm{rmax}_\gamma = a'a''$. Consider $l := g_\gamma + \mathrm{rmax}_\gamma \in M$. There is a representation $l = a'x' + a''x''$ with $x' \in M'$, $x'' \in M''$, which is easily seen to be unique. From this follows
$$\Gamma(l\,;\gamma) \;=\; \Gamma(x'\,;\gamma') \oplus \Gamma(x''\,;\gamma'')\,.$$
In particular $\mathrm{rank}\,\Gamma(l\,;\gamma) < n$. Thus $l \notin \mathrm{Sat}(\gamma)$ and $l \leq \mathrm{sat}_\gamma \leq \mathrm{vsat}_\gamma$. Equality now follows by Suppl. 6a) in Sect. 4.
    Examples where $\mathrm{rmax}_\gamma \in \mathrm{VSat}(\gamma)$ and M is not a complete intersection are provided by $\gamma = (24, 32, 40, 26, 39)$ and $\gamma = (33, 44, 55, 26, 39)$.)
b) Suppose $M \neq \mathbb{N}$. Then $\mathrm{rmax}_\gamma \leq \mathrm{con}_\gamma \leq \mathrm{vsat}_\gamma$.

(If $\mathrm{con}_\gamma > \mathrm{vsat}_\gamma$ then $\mathrm{rmax}_\gamma = \mathrm{con}_\gamma$. Thus it suffices to show $\mathrm{rmax}_\gamma \leq \mathrm{vsat}_\gamma$. Assume the opposite. Then $\mathrm{vsat}_\gamma < \mathrm{rmax}_\gamma = \mathrm{vsat}_\gamma - \mathrm{g}_\gamma$ by a), hence $\mathrm{g}_\gamma = -1$, i. e. $M = \mathbb{N}$. Contradiction. For $M = \mathbb{N}$ one has $\mathrm{rmax}_\gamma = \mathrm{con}_\gamma = \mathrm{vsat}_\gamma + 1$.)

c) Suppose $M \neq \mathbb{N}$. If $\mathrm{rmax}_\gamma > \mathrm{sat}_\gamma$ or $\mathrm{con}_\gamma > \mathrm{sat}_\gamma$, then
$$\mathrm{sat}_\gamma < \mathrm{rmax}_\gamma = \mathrm{con}_\gamma \leq \mathrm{vsat}_\gamma.$$
(The hypothesis implies $\mathrm{rmax}_\gamma = \mathrm{con}_\gamma > \mathrm{sat}_\gamma$. The inequality $\mathrm{con}_\gamma \leq \mathrm{vsat}_\gamma$ follows by b). An example of γ with $\mathrm{rmax}_\gamma > \mathrm{sat}_\gamma$ is $(6, 7, 9, 17)$. In this case $\mathrm{sat}_\gamma = 32$ and $\mathrm{rmax}_\gamma = \mathrm{con}_\gamma = \mathrm{vsat}_\gamma = 34$. Other examples are given by $\gamma = (6, 14, 21, \gamma_3)$ with $\gamma_3 = 43$ or $\gamma_3 = 37$. In these cases sat_γ, $\mathrm{rmax}_\gamma = \mathrm{con}_\gamma$, vsat_γ are $80, 86, 86$ and $73, 74, 80$ resp.)

d) The following conditions on γ are equivalent:
(1) $\mathrm{con}_\gamma \in \mathrm{VSat}(\gamma)$.
(2) $M = \mathrm{Mon}(\gamma)$ is symmetric and $\mathrm{con}_\gamma > \mathrm{g}_\gamma + \max_i \gamma_i$.
(3) $M = \mathbb{N}$ or M is a complete intersection of the following special type: There are a decomposition $\{0, \ldots, n\} = \Omega' \uplus \Omega''$ and $a', a'' \in \mathbb{N}$, $\gamma' = (\gamma_i')_{i \in \Omega'}$, $\gamma'' = (\gamma_i'')_{i \in \Omega''}$ such that $\gamma = (a'\gamma', a''\gamma'')$, $\max_i \gamma_i \leq a' + a''$ and $M = \mathrm{Mon}(a', a'')$.
(The implications (3) $\Rightarrow$ (1) and (2) $\Rightarrow$ (1) follow by direct computations. For the other implications we may assume $M \neq \mathbb{N}$. For the proof of (1) $\Rightarrow$ (3) note that $\mathrm{con}_\gamma = \mathrm{rmax}_\gamma$. Thus we may use a) and the decomposition as in 5.1. We have $\mathrm{con}_\gamma = a'a'' \geq \mathrm{g}_\gamma + \max_i \gamma_i$. Because of $a'\mathrm{g}_{\gamma'} + a''\mathrm{g}_{\gamma''} + a'a'' \notin M$ one has $\mathrm{g}_\gamma \geq a'\mathrm{g}_{\gamma'} + a''\mathrm{g}_{\gamma''} + a'a''$ and therefore $0 \geq a'\mathrm{g}_{\gamma'} + a''\mathrm{g}_{\gamma''} + \max_i \gamma_i$, which implies $\mathrm{g}_{\gamma'} = \mathrm{g}_{\gamma''} = -1$. For the proof of (2) $\Rightarrow$ (3) one uses Suppl. 5d) in Sect. 4 and assumes $\gamma_0 \leq \gamma_i \leq \gamma_n$. One has $m := \mathrm{con}_\gamma = \mathrm{g}_\gamma + \gamma_0 + \gamma_k$ and $\gamma_n < \gamma_0 + \gamma_k$ where k is any element of $\{0, \ldots, n\}$ not belonging to the connected component of 0 in $G_m(\gamma)$. Therefore it suffices to show that every $\gamma_i \neq \gamma_k$ is a multiple of γ_0. Using an induction argument it suffices to show $\gamma_i - \gamma_0 \in M$. But, if $\gamma_i - \gamma_0 \notin M$, there is an element $x \in M$ with $\gamma_i - \gamma_0 + x = \mathrm{g}_\gamma = m - \gamma_0 - \gamma_k$ and $m = \gamma_i + \gamma_k + x$. Contradiction.)

2. Assume that the numerical monoid $M = \mathrm{Mon}(\gamma) = a'M' + a''M''$ splits along Ω', Ω'', where $\Omega' \uplus \Omega'' = \{0, \ldots, n\}$. We will use the corresponding notations introduced in this section. Let v_i, $i = 0, \ldots, a'' - 1$, denote the minimal elements of M' with $v_i \equiv i$ modulo a'' and w_j, $j = 0, \ldots, a' - 1$, the minimal elements of M'' with $w_j \equiv j$ modulo a'. Then
$$u_{ij} := a'v_i + a''w_j, \quad 0 \leq i \leq a'' - 1, 0 \leq j \leq a' - 1,$$
are the minimal representatives in M for the residue classes modulo $m = a'a''$. This provides a simple proof of Theorem 5.2.

3.† Assume that the numerical monoid $M = \mathrm{Mon}(\gamma) = a'M' + a''M''$ splits along Ω', Ω'', where $\Omega' \uplus \Omega'' = \{0, \ldots, n\}$. We will use the corresponding notations introduced before.

a) As in the procedure of reduction in codimension 1 (cf. Section 3) there are canonical reduction maps $R' : \mathbb{Z} \to \mathbb{Z}$ and $R'' : \mathbb{Z} \to \mathbb{Z}$ which we define as follows: Let $m \in \mathbb{Z}$. There is a unique minimal element $s'(m) \in M''$ such that $m \equiv a''s'(m)$ modulo a' and a unique minimal element $s''(m) \in M'$ such that $m \equiv a's''(m)$ modulo a''. Then
$$R'(m) := \frac{m - a''s'(m)}{a'}, \quad R''(m) := \frac{m - a's''(m)}{a''}.$$

One proves readily (Analogous results hold for R''.):
(1) $m \in M$ if and only if $R'(m) \in M'$.
(2) $R'(m + a'z) = R'(m) + z$ for all $m, z \in \mathbb{Z}$.
(3)   If I is a fractional M-ideal, then $R'(I)$ is a fractional M'-ideal.
(4)   If I' is a fractional M'-ideal, then $I := R'^{-1}(I')$ is a fractional M-ideal with Frobenius number $g_I = a'g_{I'} + a''(g_{M''} + a')$.

b) Let us fix $\alpha' \in N_{a''}(\gamma')$, $\alpha'' \in N_{a'}(\gamma'')$ and identify $\mathbb{Z}^{n+1} = \mathbb{Z}^{\Omega'} \oplus \mathbb{Z}^{\Omega''}$. Then
$$\mathrm{Syz}(\gamma) \;=\; \mathrm{Syz}(\gamma') \oplus \mathrm{Syz}(\gamma'') \oplus \mathbb{Z}(\alpha', -\alpha'')\,.$$
There is a natural system of generators for $\mathrm{Syz}(\gamma)$ respecting this decomposition:
Let $F_j := T^{\alpha(j)} - T^{\beta(j)}$, $j \in J'$, be a system of binomial generators of the kernel of $K[T_i : i \in \Omega'] \to K[a'M']$, $\delta_j := \deg F_j = a'\delta_j'$, and $F_j := T^{\alpha(j)} - T^{\beta(j)}$, $j \in J''$, be a system of binomial generators of the kernel of $K[T_i : i \in \Omega''] \to K[a''M'']$, $\delta_j := \deg F_j = a''\delta_j''$, $J' \cap J'' = \emptyset$. Then the binomials F_j, $j \in J' \cup J''$, together with $T^{\alpha'} - T^{\alpha''}$, generate the kernel of $K[T] \to K[M]$ and
$$\mathrm{Syz}(\gamma') \;=\; \sum_{j \in J'} \mathbb{Z}(\alpha(j) - \beta(j))\,, \quad \mathrm{Syz}(\gamma'') \;=\; \sum_{j \in J''} \mathbb{Z}(\alpha(j) - \beta(j))\,.$$

c) Let $m \in M$. If $m \notin M + a'a''$, then by 4.4(2) the following holds:
$$\Gamma(m; \gamma) \;=\; \sum_{\substack{j \in J' \text{ with } m \in M + a'\delta_j'}} \mathbb{Z}(\alpha(j) - \beta(j)) \;+\; \sum_{\substack{j \in J'' \text{ with } m \in M + a''\delta_j''}} \mathbb{Z}(\alpha(j) - \beta(j))$$
$$=\; \sum_{\substack{j \in J' \text{ with } R'(m) \in M' + \delta_j'}} \mathbb{Z}(\alpha(j) - \beta(j)) \;+\; \sum_{\substack{j \in J'' \text{ with } R''(m) \in M'' + \delta_j''}} \mathbb{Z}(\alpha(j) - \beta(j))$$
$$=\; \Gamma(R'(m); \gamma') \oplus \Gamma(R''(m); \gamma'')\,.$$
If $m \in M + a'a''$, then
$$\Gamma(m; \gamma) \;=\; \Gamma(R'(m); \gamma') \oplus \Gamma(R''(m); \gamma'') \oplus \mathbb{Z}(\alpha', -\alpha'')\,.$$
As a consequence, if $m \notin M + a'a''$ then
$$\mathrm{Syz}(\gamma)/\Gamma(m; \gamma) \;=\; \mathrm{Syz}(\gamma')/\Gamma(R'(m); \gamma') \oplus \mathrm{Syz}(\gamma'')/\Gamma(R''(m); \gamma'') \oplus \mathbb{Z}(\alpha', -\alpha'')\,,$$
and if $m \in M + a'a''$ then
$$\mathrm{Syz}(\gamma)/\Gamma(m; \gamma) \;=\; \mathrm{Syz}(\gamma')/\Gamma(R'(m); \gamma') \oplus \mathrm{Syz}(\gamma'')/\Gamma(R''(m); \gamma'')\,.$$

d) From c) follows:
$$\mathrm{Sat}(\gamma) \;=\; R'^{-1}(\mathrm{Sat}(\gamma')) \cap R''^{-1}(\mathrm{Sat}(\gamma'')) \cap (M + a'a'')\,,$$
$$\mathrm{VSat}(\gamma) \;=\; R'^{-1}(\mathrm{VSat}(\gamma')) \cap R''^{-1}(\mathrm{VSat}(\gamma'')) \cap (M + a'a'')\,.$$

The Frobenius numbers sat_γ and vsat_γ are therefore the maxima of the Frobenius numbers of the M-ideals on the right side of the equations above, which can be written down using a)(4). (Examples show that none of the ideals in the intersection formulas can be neglected, for instance, consider $\gamma' = (3, 8)$, $\gamma'' = (5, 7)$ and (a', a'') in the three cases $(5, 3)$, $(5, 6)$, $(36, 25)$.)

II Regular Sequences

6 Regular Sequences and Complete Intersections

The concept of a complete intersection is an old one. Nevertheless there are some terms and details not generally agreed upon. We therefore describe part of the set-up in the manner we are going to use.

Roughly speaking, complete intersections are defined by ideals generated by regular sequences or prime sequences, as they are sometimes called. There are ambiguities about the concept of a regular sequence, though. Thus some remarks on our conventions are in order.

Consider a noetherian ring R and a sequence $f_1, \ldots, f_r$ of elements in R. We will call $f_1, \ldots, f_r$ a s t r o n g l y   r e g u l a r   s e q u e n c e in R, if $R \neq Rf_1 + \cdots + Rf_r$ and if for every $i = 1, \ldots, r$ the element f_i is a non-zero-divisor on $R/(Rf_1 + \cdots + Rf_{i-1})$. The sequence $f_1, \ldots, f_r$ is called a r e g u l a r   s e q u e n c e in R, if for every prime ideal $\mathfrak{p}$ in R containing $f_1, \ldots, f_r$ the sequence $f_1, \ldots, f_r$ is strongly regular in the local ring $R_{\mathfrak{p}}$. It suffices to check this condition for all maximal ideals in R containing $f_1, \ldots, f_r$. Of course, a strongly regular sequence is a regular sequence.

More generally, strongly regular sequences and regular sequences can be defined for a finitely generated R-module V. How to do this is obvious in the case of a strongly regular sequence. A sequence $f_1, \ldots, f_r$ in R is a r e g u l a r   s e q u e n c e for V, if for every prime ideal $\mathfrak{p}$ in the support of $V/(f_1V + \cdots + f_rV)$ the sequence $f_1, \ldots, f_r$ is strongly regular for $V_{\mathfrak{p}}$.

The difference between regular and strongly regular sequences is well illustrated by the following statement:

6.1 *Let V be a finite module over the noetherian ring R and let $f_1, \ldots, f_r$ be a sequence in R with $V \neq f_1V + \cdots + f_rV$. Then the following conditions are equivalent:*

(1)   *$f_1, \ldots, f_r$ is a strongly regular sequence for V.*

(2)   *For $s = 1, \ldots, r$ the sequence $f_1, \ldots, f_s$ is a regular sequence for V.*

Proof. (2) follows from (1) in a trivial way. To prove the converse we have to show that for every prime ideal $\mathfrak{p}$ of R and every $i = 1, \ldots, r$ the element f_i is a non-zero-divisor on $W := V_{\mathfrak{p}}/(f_1V_{\mathfrak{p}} + \cdots + f_{i-1}V_{\mathfrak{p}})$. This is clear by our assumption (2) in case $\mathfrak{p}$ belongs to the support of $V/(f_1V + \cdots + f_iV)$. If $V_{\mathfrak{p}} = f_1V_{\mathfrak{p}} + \cdots + f_iV_{\mathfrak{p}}$ then multiplication with f_i on W is surjective and hence even bijective.   □

If $f_1, \ldots, f_r$ is a regular sequence for V, then for every permutation $\pi \in \mathfrak{S}_r$ the sequence $f_{\pi 1}, \ldots, f_{\pi r}$ is also regular for V. For strongly regular sequences 6.1 implies: Let $V \neq f_1 V + \cdots + f_r V$. Then the sequence $f_{\pi 1}, \ldots, f_{\pi r}$ is strongly regular for every $\pi \in \mathfrak{S}_r$ if and only if *all* subsequences of $f_1, \ldots, f_r$ are regular.

A regular sequence could be rightly called a locally regular sequence. However, this seems to be a tedious notation the use of which does not reflect the fact that regular sequences occur in a basic way, whereas strongly regular sequences appear mostly in special circumstances. Therefore, from now on we will use, without further notice, the concepts regular/strongly regular as defined above.

Throughout this section, let B always be a (commutative) finitely generated algebra over the noetherian ring A.

Definition B is called a l o c a l l y c o m p l e t e i n t e r s e c t i o n of relative dimension s over A if B is A-flat and if for every maximal ideal $\mathfrak{M}$ in B the local ring
$$B_{\mathfrak{M}}/(A \cap \mathfrak{M})B_{\mathfrak{M}}$$
is a complete intersection of (Krull) dimension s in the absolute sense of local algebra.

Thereby, a noetherian local ring R, which is isomorphic to a residue class ring $S/\mathfrak{a}$ of a regular local ring S, is called a c o m p l e t e i n t e r s e c t i o n (in the a b s o l u t e sense) if the ideal $\mathfrak{a}$ is generated by a regular sequence in S. Let us observe that the ring $R := B_{\mathfrak{M}}/(A \cap \mathfrak{M})B_{\mathfrak{M}}$ in the definition above is indeed a homomorphic image of a regular local ring: Since we assume that B is a finitely generated A-algebra, there is a representation $B = A[X_1, \ldots, X_n]/\mathfrak{A}$. Thus the ring R is a homomorphic image of the regular local ring $S := (A_{\mathfrak{p}}/\mathfrak{p}A_{\mathfrak{p}})[X_1, \ldots, X_n]_{\mathfrak{M}}$, $\mathfrak{p} := A \cap \mathfrak{M}$. For a prime ideal $\mathfrak{p}$ in A, the algebra $B_{\mathfrak{p}}/\mathfrak{p}B_{\mathfrak{p}} = \kappa(\mathfrak{p}) \otimes_A B$ over the residue field
$$\kappa(\mathfrak{p}) := A_{\mathfrak{p}}/\mathfrak{p}A_{\mathfrak{p}}$$
of $A_{\mathfrak{p}}$ is called the f i b r e a l g e b r a or the f i b r e of B over $\mathfrak{p}$. For further information about complete intersections in the absolute sense see Supplement 5.

There are the following characterizations of locally complete intersections:

6.2 Theorem *For $B = A[X_1, \ldots, X_n]/\mathfrak{A}$ and a nonnegative integer s the following conditions are equivalent:*
(1) *B is a locally complete intersection of relative dimension s over A.*
(2) *B is A-flat and for every $\mathfrak{P} \in \operatorname{Spec} B = \mathrm{V}(\mathfrak{A}) \subseteq \operatorname{Spec} A[X_1, \ldots, X_n]$ the ideal $\mathfrak{A}_{\mathfrak{P}}$ in $A[X_1, \ldots, X_n]_{\mathfrak{P}}$ is generated by a regular sequence of length $n - s$.*

(3) B is A-flat and the conormal module $\mathfrak{A}/\mathfrak{A}^2$ is a projective B-module of rank $n - s$.

(4)   For every $\mathfrak{M} \in \operatorname{Spm} B \subseteq V(\mathfrak{A}) \subseteq \operatorname{Spec} A[X_1,\ldots,X_n]$ the local ring $B_\mathfrak{M}/(A \cap \mathfrak{M})B_\mathfrak{M}$ has dimension s and the ideal $\mathfrak{A}_\mathfrak{M}$ is generated by $n - s$ elements.

For the proof of 6.2 we will use the following lemmata. The first one is essentially due to M. Nagata (cf. [38], §1):

6.3 Lemma (Nagata) Let $A \to P$ be a local homomorphism of local noetherian rings and V a finite P-module which is flat as an A-module. Let $f_1,\ldots,f_r$ be elements of $\mathfrak{m}_P$ forming a regular sequence for $V/\mathfrak{m}_A V$. Then $f_1,\ldots,f_r$ is a regular sequence for $V/\mathfrak{a}V$, $\mathfrak{a} \subseteq A$ an arbitrary ideal. Moreover, the module $V/(f_1,\ldots,f_r)V$ is A-flat.

Proof of 6.3.   By induction on r we may assume $r = 1$. Let $f := f_1$.

First we show that f is a non-zero-divisor on $V/\mathfrak{a}V$. Using noetherian induction we may assume, as induction hypothesis, that multiplication by f is injective on all modules $V/\mathfrak{a}V$ where $\mathfrak{a}$ runs through the ideals $\neq 0$ of A, and have to prove that multiplication by f on V itself is injective.

First we consider the case where there is an ideal $\mathfrak{a} \cong A/\mathfrak{m}_A$ in A. By A-flatness of V we have a canonical commutative diagram

$$
\begin{array}{ccccccccc}
0 & \longrightarrow & \mathfrak{a} \otimes_A V & \longrightarrow & V & \longrightarrow & V/\mathfrak{a}V & \longrightarrow & 0 \\
 & & \downarrow{\scriptstyle f} & & \downarrow{\scriptstyle f} & & \downarrow{\scriptstyle f} & & \\
0 & \longrightarrow & \mathfrak{a} \otimes_A V & \longrightarrow & V & \longrightarrow & V/\mathfrak{a}V & \longrightarrow & 0
\end{array}
$$

with exact rows. By induction hypothesis the outer homomorphisms are injective. Thus $V \xrightarrow{f} V$ is injective, too.

Otherwise let $a \in \mathfrak{m}_A$ be a non-zero-divisor in A. Again by flatness the canonical commutative diagram

$$
\begin{array}{ccccccccc}
0 & \longrightarrow & V & \xrightarrow{a} & V & \longrightarrow & V/aV & \longrightarrow & 0 \\
 & & \downarrow{\scriptstyle f} & & \downarrow{\scriptstyle f} & & \downarrow{\scriptstyle f} & & \\
0 & \longrightarrow & V & \xrightarrow{a} & V & \longrightarrow & V/aV & \longrightarrow & 0
\end{array}
$$

has exact rows. By induction hypothesis multiplication by f on V/aV is injective, which implies that multiplication by a on the kernel K of multiplication by f on V is surjective: $K = aK$. By Krull-Nakayama's lemma $K = 0$.

To prove finally that V/fV is A-flat we just have to show that the module $\operatorname{Tor}_1^A(A/\mathfrak{a}, V/fV)$ vanishes for every ideal $\mathfrak{a}$ in A. By flatness of V this means that multiplication by f is injective on $(A/\mathfrak{a}) \otimes V = V/\mathfrak{a}V$ which however we proved above.   □

6.4 Lemma (Vasconcelos) *Let R be a local noetherian ring and $\mathfrak{a} \subseteq \mathfrak{m}_R$ an ideal such that $S := R/\mathfrak{a}$ has finite projective dimension over R. Assume that $\mathfrak{a}/\mathfrak{a}^2$ is S-free of rank r. Then $\mathfrak{a}$ can be generated by a regular sequence (of length r).*

Proof by induction on r (cf. [46]). Assume $r > 0$ and let Q be the total quotient ring of R. Then $Q/\mathfrak{a}Q$ has a finite free resolution over Q and is therefore Q-free (Lemma of Auslander and Buchsbaum). This implies $\mathfrak{a}Q = Q$, and $\mathfrak{a}$ therefore contains a non-zero-divisor. Obviously then there is a $f \in \mathfrak{a}, f \notin \mathfrak{m}\mathfrak{a}$, which is a non-zero-divisor of R. Let $\bar{R} := R/Rf$ and $\bar{\mathfrak{a}} := \mathfrak{a}/Rf$. Then $\bar{\mathfrak{a}}/\bar{\mathfrak{a}}^2$ is free over $\bar{S} = \bar{R}/\bar{\mathfrak{a}} = S$ with basis $\bar{f}_2, \ldots, \bar{f}_r$ where $f, f_2, \ldots, f_r$ is a minimal set of generators for $\mathfrak{a}$.

To finish the induction we have to show that S has finite projective dimension over $\bar{R}$. But $\mathfrak{a}/f\mathfrak{a}$ has finite projective dimension over $\bar{R}$ (because f is a non-zero-divisor for $\mathfrak{a}$ and R), and S is a direct summand of $\mathfrak{a}/f\mathfrak{a}$ because $S \cong fR/f\mathfrak{a} \subseteq \mathfrak{a}/f\mathfrak{a}$ is mapped by the canonical projection $\mathfrak{a}/f\mathfrak{a} \to \mathfrak{a}/\mathfrak{a}^2$ to the direct summand $S\bar{f} \cong S$.   □

Proof of 6.2. (1) implies (2): It suffices to prove (2) for maximal ideals $\mathfrak{M} \in V(\mathfrak{A})$. Set $\mathfrak{p} := A \cap \mathfrak{M}$, $k := A_\mathfrak{p}/\mathfrak{p}A_\mathfrak{p}$ and $X = (X_1, \ldots, X_n)$. Because $B_\mathfrak{M}$ is flat over $A_\mathfrak{p}$ we get from the exact sequence $0 \to \mathfrak{A}_\mathfrak{M} \to A[X]_\mathfrak{M} \to B_\mathfrak{M} \to 0$ the exact sequence

$$0 \longrightarrow \mathfrak{A}_\mathfrak{M}/\mathfrak{p}\mathfrak{A}_\mathfrak{M} \longrightarrow k[X]_\mathfrak{M} \longrightarrow B_\mathfrak{M}/\mathfrak{p}B_\mathfrak{M} \longrightarrow 0.$$

Since $B_\mathfrak{M}/\mathfrak{p}B_\mathfrak{M}$ is a complete intersection of dimension s there are functions $f_1, \ldots, f_{n-s}$ in $\mathfrak{A}$ which generate $\mathfrak{A}_\mathfrak{M}/\mathfrak{p}\mathfrak{A}_\mathfrak{M}$ and form a regular sequence in

$$k[X]_\mathfrak{M} = A_\mathfrak{p}[X]_\mathfrak{M}/\mathfrak{p}A_\mathfrak{p}[X]_\mathfrak{M}.$$

By Krull-Nakayama's Lemma $f_1, \ldots, f_{n-s}$ generate $\mathfrak{A}_\mathfrak{M}$ and by Lemma 6.3 they form a regular sequence in $A[X]_\mathfrak{M}$.

(2) implies (3) because for any strongly regular sequence $c_1, \ldots, c_r$ in a commutative ring C the module $\mathfrak{c}/\mathfrak{c}^2$ is free of rank r over $C/\mathfrak{c}$, $\mathfrak{c} := Cc_1 + \cdots + Cc_r$.

(3) implies (4): Let $\mathfrak{p} = A \cap \mathfrak{M}$ and $k = A_\mathfrak{p}/\mathfrak{p}A_\mathfrak{p}$ as above. Then for the kernel $\mathfrak{A}'$ of the canonical homomorphism $k[X]_\mathfrak{M} \to B_\mathfrak{M}/\mathfrak{p}B_\mathfrak{M}$ the $(B_\mathfrak{M}/\mathfrak{p}B_\mathfrak{M})$-module $\mathfrak{A}'/\mathfrak{A}'^2 = k \otimes (\mathfrak{A}_\mathfrak{M}/\mathfrak{A}_\mathfrak{M}^2)$ (cf. Lemma 6.7) is free of rank $n - s$. By Lemma 6.4 $\mathfrak{A}'$ is generated by a regular sequence in $k[X]_\mathfrak{M}$ and $B_\mathfrak{M}/\mathfrak{p}B_\mathfrak{M}$ has dimension s.

(4) implies (1): We have only to show that $B_\mathfrak{M} = A_\mathfrak{p}[X]_\mathfrak{M}/\mathfrak{A}_\mathfrak{M}$ is flat over $A_\mathfrak{p}$. This is a special case of Lemma 6.3.   □

The conormal module $\mathfrak{A}/\mathfrak{A}^2$ of a representation $B = A[X_1, \ldots, X_n]/\mathfrak{A}$ of a locally complete intersection determines an invariant of B over A. To define this we notice:

6.5 Lemma *Let $B = A[X_1, \ldots, X_n]/\mathfrak{A}$ and let $y_1, \ldots, y_m$ be elements in B with representatives $G_1, \ldots, G_m \in A[X_1, \ldots, X_n]$. Then the kernel $\mathfrak{C}$ of the homomorphism $A[X_1, \ldots, X_n, Y_1, \ldots, Y_m] \to B$ which extends $A[X_1, \ldots, X_n] \to B$ by $Y_j \mapsto y_j$, $j = 1, \ldots, m$, is generated by $\mathfrak{A}$ and $Y_j - G_j$, $j = 1, \ldots, m$. Furthermore there is a B-isomorphism*

$$\mathfrak{C}/\mathfrak{C}^2 = (\mathfrak{A}/\mathfrak{A}^2) \oplus B^m.$$

Proof. We may assume $m = 1$. Clearly $\mathfrak{C}$ is generated by $\mathfrak{A}$, $Z := Y_1 - G_1$ and $\mathfrak{C}^2$ is generated by $\mathfrak{A}^2, Z\mathfrak{A}, Z^2$. We have $\mathfrak{C} \equiv \mathfrak{A} \oplus A[X_1, \ldots, X_n]z$, $\mathfrak{C}^2 \equiv \mathfrak{A}^2 \oplus \mathfrak{A}z$ modulo Z^2. Thus $\mathfrak{C}/\mathfrak{C}^2 = (\mathfrak{A}/\mathfrak{A}^2) \oplus (A[X_1, \ldots, X_n]/\mathfrak{A})$. $\quad\square$

6.6 Corollary *For two representations $B = A[X_1, \ldots, X_n]/\mathfrak{A}$ and $B = A[Y_1, \ldots, Y_m]/\mathfrak{B}$ of any finitely generated A-algebra B the corresponding conormal modules are stably equivalent, more precisely:*

$$(\mathfrak{A}/\mathfrak{A}^2) \oplus B^m = (\mathfrak{B}/\mathfrak{B}^2) \oplus B^n.$$

Proof. $B = A[x_1, \ldots, x_n] = A[x_1, \ldots, x_n, y_1, \ldots, y_m] = A[y_1, \ldots, y_m]$. Now use 6.5. $\quad\square$

Let B be a locally complete intersection. By 6.2 and 6.6 the conormal module $\mathfrak{A}/\mathfrak{A}^2$ of any representation $B = A[X_1, \ldots, X_n]/\mathfrak{A}$ determines a well-defined class in the projective class group $\tilde{K}_0(B)$, namely

$$c_A(B) := [\mathrm{Hom}_B(\mathfrak{A}/\mathfrak{A}^2, B)] \in \tilde{K}_0(B)$$

which we call the **c a n o n i c a l   c l a s s** of B over A.

To discuss base changes $A \to A'$ we start with the following observation:

6.7 *Assume $B = A[X_1, \ldots, X_n]/\mathfrak{A}$ to be flat over A and let $A \to A'$ be any base change. Then $B' := A' \otimes_A B = A'[X_1, \ldots, X_n]/\mathfrak{A}'$, where $\mathfrak{A}' := \mathfrak{A}A'[X_1, \ldots, X_n]$, and*

$$\mathfrak{A}'/\mathfrak{A}'^2 = A' \otimes_A (\mathfrak{A}/\mathfrak{A}^2) = B' \otimes_B (\mathfrak{A}/\mathfrak{A}^2).$$

Proof. Because B is A-flat the canonical homomorphism $A' \otimes_A \mathfrak{A} \to \mathfrak{A}'$ is bijective. Since $A' \otimes_A \mathfrak{A}^2 \to \mathfrak{A}'^2$ is surjective we get $\mathfrak{A}'/\mathfrak{A}'^2 \cong A' \otimes_A (\mathfrak{A}/\mathfrak{A}^2)$ as asserted. $\quad\square$

By 6.2(3) and 6.7 the following is obvious.

6.8 Proposition *Let B be a finitely generated flat A-algebra, $A \to A'$ a base change and B' the A'-algebra $A' \otimes_A B$.*
(1) If B is a locally complete intersection (of relative dimension s) over A then so is B' over A'. In this case

$$c_{A'}(B') = B' \otimes_B c_A(B).$$

(2)   If $A \to A'$ is faithfully flat and B' is a locally complete intersection (of relative dimension s) over A' then so is B over A.

Proposition 6.8 can be used to prove a point criterion for locally complete intersections:

6.9 Proposition $B = A[X_1, \ldots, X_n]/\mathfrak{A}$ is a locally complete intersection of relative dimension s if and only if B is A-flat and if for every $\mathfrak{p} \in \operatorname{Spec} A$ the fibre

$$B_\mathfrak{p}/\mathfrak{p}B_\mathfrak{p} \;=\; \kappa(\mathfrak{p})[X_1, \ldots, X_n]/(\mathfrak{A})$$

is a locally complete intersection over $\kappa(\mathfrak{p}) = A_\mathfrak{p}/\mathfrak{p}A_\mathfrak{p}$ of relative dimension s. (($\mathfrak{A}$) denotes the extension of $\mathfrak{A}$ in $\kappa(\mathfrak{p})[X_1, \ldots, X_n]$.)

Proof. Assume B to be a locally complete intersection of relative dimension s. Then so are all fibres by 6.8. Conversely, assume the fibres to be locally complete intersections of relative dimension s and B to be A-flat. Consider $\mathfrak{M} \in \operatorname{Spm} B$ and $\mathfrak{p} := A \cap \mathfrak{M}$. Then $B_\mathfrak{M}/\mathfrak{p}B_\mathfrak{M}$ is by assumption on the fibres a complete intersection of dimension s. $\qquad\square$

The case, where the canonical class of a locally complete intersection is trivial, is of special interest.

Definition A locally complete intersection B over the noetherian ring A (of relative dimension s) is called a complete intersection (of relative dimension s) if its canonical class $c_A(B)$ is trivial, i.e. if the conormal module $\mathfrak{A}/\mathfrak{A}^2$ of one and then of any representation $B = A[X_1, \ldots, X_n]/\mathfrak{A}$ is stably free.

By the definitions the following proposition is obvious.

6.10 Proposition Let $B = A[X_1, \ldots, X_n]/\mathfrak{A}$ be a locally complete intersection of relative dimension s. If the ideal $\mathfrak{A}$ is generated by $n - s$ polynomials $F_1, \ldots, F_{n-s}$, then $\mathfrak{A}/\mathfrak{A}^2$ is free as a B-module with basis $F_1 + \mathfrak{A}^2, \ldots, F_{n-s} + \mathfrak{A}^2$, in particular, B is a complete intersection over A.

A more precise version of 6.10 is part of the following theorem.

6.11 Theorem Let B be a finitely generated algebra over the noetherian ring A. The following are equivalent:
(1) B is a complete intersection of relative dimension s over A.
(2) B is A-flat and there is a representation $B = A[X_1, \ldots, X_n]/\mathfrak{A}$ such that the conormal module $\mathfrak{A}/\mathfrak{A}^2$ is B-free of rank $n - s$.
(3) B is A-flat and there are a representation $B = A[X_1, \ldots, X_n]/\mathfrak{A}$ and polynomials $F_1, \ldots, F_{n-s} \in \mathfrak{A}$ such that for every $\mathfrak{P} \in \mathrm{V}(\mathfrak{A}) = \operatorname{Spec} B$ the following holds: $F_1, \ldots, F_{n-s}$ generate $\mathfrak{A}_\mathfrak{P}$ and form a regular sequence in $A[X_1, \ldots, X_n]_\mathfrak{P}$.

(4)   *B is A-flat and there is a representation $B = A[Y_1, \ldots, Y_m]/\mathfrak{B}$ where $\mathfrak{B}$ is generated by a regular sequence $G_1, \ldots, G_{m-s}$ in $A[Y_1, \ldots, Y_m]$.*

(5)   *There is a representation $B = A[Y_1, \ldots, Y_m]/(G_1, \ldots, G_{m-s})$ and all the non-zero fibres of B over A have dimension s.*

If condition (2) holds then (3) is fulfilled for polynomials $F_1, \ldots, F_{n-s} \in \mathfrak{A}$ the residue classes of which form a B-basis of $\mathfrak{A}/\mathfrak{A}^2$. Conversely, if (3) holds then the residue classes of $F_1, \ldots, F_{n-s}$ form a B-basis of $\mathfrak{A}/\mathfrak{A}^2$.

If (3) holds then there is an element $y \in B$ such that the kernel $\mathfrak{B}$ of the homomorphism $A[X_1, \ldots, X_n, Y] \to B$, which extends $A[X_1, \ldots, X_n] \to B$ by $Y \mapsto y$, is generated by polynomials $F_1, \ldots, F_{n-s} \in \mathfrak{A}$ and one further polynomial $G = 1 - FY$ with $F \in A[X_1, \ldots, X_n]$.

Proof. (1) and (2) are equivalent by 6.5.

(2) implies (3): Let $F_1, \ldots, F_{n-s}$ be any elements of $\mathfrak{A}$ such that their residue classes form a basis of $\mathfrak{A}/\mathfrak{A}^2$. Then $F_1, \ldots, F_{n-s}$ generate $\mathfrak{A}_\mathfrak{P}$ for every $\mathfrak{P} \in V(\mathfrak{A})$ by Krull-Nakayama's lemma. By 6.2(2) they form a regular sequence in $A[X_1, \ldots, X_n]_\mathfrak{P}$.

(3) implies (2): The residue classes of $F_1, \ldots, F_{n-s}$ form a $B_\mathfrak{P}$-basis of $\mathfrak{A}_\mathfrak{P}/\mathfrak{A}_\mathfrak{P}^2$ for every $\mathfrak{P} \in \operatorname{Spec} B$ and therefore a B-basis for $\mathfrak{A}/\mathfrak{A}^2$.

(3) implies (4): Let $\mathfrak{A}' \subseteq \mathfrak{A}$ be the ideal generated by $F_1, \ldots, F_{n-s}$ and consider the ring $B' := A[X_1, \ldots, X_n]/\mathfrak{A}'$. The kernel $\mathfrak{A}/\mathfrak{A}'$ of the canonical projection $B' \to B$ is idempotent because of $\mathfrak{A}^2 + \mathfrak{A}' = \mathfrak{A}$ and is therefore generated by an idempotent element $e' \in B'$. The induced homomorphism $B'_{1-e'} \to B$ is bijective. Now let $F \in A[X_1, \ldots, X_n]$ be a polynomial with residue class $1 - e'$ in B'. Then $A[X_1, \ldots, X_n, Y]/(F_1, \ldots, F_{n-s}, 1 - FY) = B'_{1-e'} = B$. Condition (3) is weaker than (4).

(4) and (1) imply (5), using 6.9.

(5) implies (1): We use the criterion given by (4) in Theorem 6.2. Let $\mathfrak{M} \in \operatorname{Spm} B \subseteq V(G_1, \ldots, G_{m-s})$, $\mathfrak{p} := A \cap \mathfrak{M}$ and $\kappa(\mathfrak{p}) := A_\mathfrak{p}/\mathfrak{p} A_\mathfrak{p}$. Then $\mathfrak{M}$ is a maximal ideal in the fibre $B_\mathfrak{p}/\mathfrak{p} B_\mathfrak{p} = \kappa(\mathfrak{p})[Y_1, \ldots, Y_m]/(G_1, \ldots, G_{m-s})$, and therefore the localization $B_\mathfrak{M}/\mathfrak{p} B_\mathfrak{M}$ has dimension $\geq s$ and hence $= s$ because the dimension of the fibre is assumed to be s.

The proof also yields the additional assertions of 6.11.   □

Let B be a complete intersection of relative dimension s over A. Then polynomials $F_1, \ldots, F_{n-s}$ as in 6.11(3) are called l o c a l l y   d e f i n i n g f u n c t i o n s for B over A, whereas polynomials $G_1, \ldots, G_{m-s}$ as in 6.11(4) are called g l o b a l l y   d e f i n i n g   f u n c t i o n s for B over A.

Globally defining functions for complete intersections always form regular sequences, locally defining functions not necessarily so.

6.12 Corollary *Let $B = A[X_1, \ldots, X_n]/\mathfrak{A}$ be a locally complete intersection of relative dimension s over A and $F_1, \ldots, F_{n-s} \in \mathfrak{A}$.*

(1)   If $F_1, \ldots, F_{n-s}$ generate $\mathfrak{A}_{\mathfrak{P}}$ for every $\mathfrak{P} \in V(\mathfrak{A})$, then B is a complete intersection over A and $F_1, \ldots, F_{n-s}$ is a system of locally defining functions.

(2)   If $F_1, \ldots, F_{n-s}$ even generate $\mathfrak{A}$, then B is a complete intersection over A and $F_1, \ldots, F_{n-s}$ is a system of globally defining functions.

Proof. In both cases $F_1, \ldots, F_{n-s}$ generate the projective B-module $\mathfrak{A}/\mathfrak{A}^2$ of rank $n - s$ (in case (1), because they do so locally) and hence form a B-basis. Thus B is a complete intersection over A. The other assertions follow from Theorem 6.11.   □

6.13 Corollary   If $B = A[X_1, \ldots, X_n]/(F_1, \ldots, F_n)$ is an A-algebra with finite fibres, then B is a complete intersection (of relative dimension 0) over A with globally defining functions $F_1, \ldots, F_n$.

Naturally, the finiteness condition in Corollary 6.13 is satisfied if B is finite over A.

However, complete intersections of relative dimension 0 need not be finite even if they are faithful. To have an example start with a finite faithful complete intersection B over A and a nonunit $f \in B$ such that the open set $D(f)$ intersects with every fibre of $\operatorname{Spec} B$ over $\operatorname{Spec} A$. Then $C := B_f$ is a faithful complete intersection over A of relative dimension 0 but in general not finite. For instance, $C = \mathbb{Z}[X]/(2X^2 + X + 1)$ is such an example over the ring $\mathbb{Z}$ of integers.

The A-algebra $B = A[X_1, \ldots, X_n]/(F_1, \ldots, F_n)$ is a *finite* complete intersection if and only if the fibres $B_{\mathfrak{p}}/\mathfrak{p} B_{\mathfrak{p}}$, $\mathfrak{p} \in \operatorname{Spec} A$, are finite of locally constant rank in $\operatorname{Spec} A$. This is a special case of the following lemma proven in [3], Lemma 2.7:

6.14 Lemma   A finitely generated flat algebra B over a noetherian ring A is finite if and only if the fibres $B_{\mathfrak{p}}/\mathfrak{p} B_{\mathfrak{p}}$, $\mathfrak{p} \in \operatorname{Spec} A$, are finite of locally constant rank in $\operatorname{Spec} A$.

Let B be a locally complete intersection over A which is finite over A, and let $B = A[X_1, \ldots, X_n]/\mathfrak{A}$ be a representation. Then the determinant

$$\det c_A(B) \; = \; [\bigwedge^n \operatorname{Hom}_B(\mathfrak{A}/\mathfrak{A}^2, B)] \; \in \; \operatorname{Pic} B$$

of the canonical class $c_A(B)$ is just the class of the r e l a t i v e   d u a l i z i n g m o d u l e $E := \operatorname{Hom}_A(B, A)$ in $\operatorname{Pic} B$:

$$\det c_A(B) \; = \; [E]$$

([39], Lemma 4.7), see Supplement 8. This is useful if the homomorphism

$$\det : \; \tilde{K}_0(B) \longrightarrow \operatorname{Pic} B$$

is bijective, which is, for instance, the case if the Krull dimension of B is ≤ 1. Then B is a complete intersection over A if and only if E is a free B-module, i. e. if B is a Frobenius algebra over A.

In a number theoretical context let us consider a finite extension $A \subseteq B$ of Dedekind domains for which the extension $Q(A) \subseteq Q(B)$ of their quotient fields is separable. Then B is a locally complete intersection over A and E is isomorphic to Dedekind's complementary module $\mathfrak{C}_A(B)$, the dual of which is Dedekind's different $\mathfrak{D}_A(B)$. Therefore, cf. [39], (4.8):

6.15 Theorem *Let $A \subseteq B$ be a finite extension of Dedekind domains for which $Q(A) \subseteq Q(B)$ is separable. Then B is a complete intersection over A if and only if $\mathfrak{D}_A(B)$ is a principal ideal.*

Supplements

1. Let $B_1, \ldots, B_m$ be finitely generated locally complete intersections over A of relative dimensions $s_1, \ldots, s_m$. Then the tensor product $B = B_1 \otimes \cdots \otimes B_m$ is a locally complete intersection over A of relative dimension $s := s_1 + \cdots + s_m$, and its canonical class is
$$c_A(B) = \sum_{i=1}^{m} B_1 \otimes \cdots \otimes B_{i-1} \otimes c_A(B_i) \otimes B_{i+1} \otimes \cdots \otimes B_m .$$
If $B_1, \ldots, B_m$ are complete intersections then B is also a complete intersection.

2. Let $A \to B$ and $B \to C$ be locally complete intersections of finite type and relative dimensions s, t. Then $A \to C$ is a locally complete intersection of relative dimension $s + t$ and its canonical class is
$$c_A(C) = C \otimes_B c_A(B) + c_B(C) .$$
(For representations $B = A[X_1, \ldots, X_m]/\mathfrak{A}$, $C = B[Y_1, \ldots, Y_n]/\mathfrak{B}$ and the derived representation $C = A[X_1, \ldots, X_m, Y_1, \ldots, Y_n]/\mathfrak{C}$ one has a canonical exact sequence of C-modules
$$0 \longrightarrow C \otimes_B (\mathfrak{A}/\mathfrak{A}^2) \longrightarrow \mathfrak{C}/\mathfrak{C}^2 \longrightarrow \mathfrak{B}/\mathfrak{B}^2 \longrightarrow 0 .)$$

3. A finite A-algebra B is a locally complete intersection over A if and only if $B(U)$ is a complete intersection over $A(U) = A(U_i : i \in I)$, provided $I \neq \emptyset$.

4. Let $F \in A[X_1, \ldots, X_n]$. Then $B := A[X]/(F)$ is a complete intersection of relative dimension $n - 1$ (with globally defining function F) if and only if F is a primitive polynomial. (In contrast to the homogeneous case, cf. 7.8, the result cannot be generalized to the case of lower relative dimension as for instance is shown by the example: $F_1 = (aX_1 + 1)X_2$, $F_2 = (aX_1 + 1)X_3$, where $a \neq 0$ is a non-zero-divisor in the maximal ideal $\mathfrak{m}$ of a local ring A; these polynomials define a complete intersection of relative dimension 1 modulo $\mathfrak{m}$ but not over A.)

In case $n = 1$ the algebra $B = A[X]/(F) = A[x]$ is a complete intersection of relative dimension 0 over A and hence an algebra with finite fibres if and only if F is primitive. If in addition the spectrum of A is connected, B is a finite complete intersection if and only if the coefficients of $F = \sum_{i=0}^{m} a_i X^i$ have the following properties: There is an index r such that a_r is a unit in A and $a_{r+1}, \ldots, a_m$ are nilpotent. (In this case B is free of rank r over A with A-basis $1, x, \ldots, x^{r-1}$.)

5.† In this supplement we discuss the concept of a complete intersection in the absolute sense of local algebra, following [35].

For an arbitrary local ring R we denote by $\mathfrak{m}_R$ its maximal ideal and by k_R the residue field $R/\mathfrak{m}_R$. We assume that all rings considered are noetherian.

By definition, the e m b e d d i n g   d i m e n s i o n embd R of R is the minimal number of generators of $\mathfrak{m}_R$, i.e. embd $R = \dim_{k_R} \mathfrak{m}_R/\mathfrak{m}_R^2$. R is regular if and only if embd $R = \dim R$.

Now let S be a regular local ring, $\mathfrak{a}$ an ideal in S with $\mathfrak{a} \subseteq \mathfrak{m}_S$ and $R := S/\mathfrak{a}$. We set $k := k_S = k_R$.

a) The embedding dimension of R is equal to $\dim_k \mathfrak{m}_S/(\mathfrak{a} + \mathfrak{m}_S^2)$. If $f_1, \ldots, f_r$ is a minimal system of generators of $\mathfrak{a}$, then there exists a subsequence $f_{i_1}, \ldots, f_{i_s}$ which induces a k-base of $(\mathfrak{a} + \mathfrak{m}_S^2)/\mathfrak{m}_S^2$ and is in particular part of a minimal system of generators of $\mathfrak{m}_S$. As a consequence, $R = \bar{S}/\bar{\mathfrak{a}}$, $\bar{S} := S/(f_{i_1}, \ldots, f_{i_s})$, $\bar{\mathfrak{a}} := \mathfrak{a}\bar{S} = \mathfrak{a}/(f_{i_1}, \ldots, f_{i_s})$, where $\bar{S}$ is regular, $\bar{\mathfrak{a}} \subseteq \mathfrak{m}_{\bar{S}}^2$ and $\dim \bar{S} = $ embd R.

b) Let r denote the minimal number of generators of $\mathfrak{a}$. Then the non-negative number
$$\text{dev } R := r - \text{ht }\mathfrak{a} = r - (\dim S - \dim R)$$
is, as we shall see in c), independent of the representation $R = S/\mathfrak{a}$. It is called the d e v i a t i o n of the local ring R. By part a),
$$\text{dev } R = \dim_k(\mathfrak{a} \cap \mathfrak{m}_S^2)/\mathfrak{m}_S\mathfrak{a} - (\text{embd } R - \dim R)$$
$$= \dim_k \bar{\mathfrak{a}}/\mathfrak{m}_{\bar{S}}\bar{\mathfrak{a}} - (\text{embd } R - \dim R).$$

c) Let $b_1 = b_1(R) := $ embd R and let $b_2 = b_2(R)$ denote the minimal number of generators of the syzygy module $\text{Syz}_R \mathfrak{m}_R$ with respect to a minimal system of generators of $\mathfrak{m}_R$. (b_1 and b_2 are obviously invariants of R.) Then
$$b_2 = \binom{b_1}{2} + \dim_k \bar{\mathfrak{a}}/\mathfrak{m}_{\bar{S}}\bar{\mathfrak{a}}$$
and hence
$$\text{dev } R = \dim R + b_2 - \binom{b_1+1}{2}.$$
(For the proof one can assume $\mathfrak{a} \subseteq \mathfrak{m}_S^2$ and $n := b_1 = \dim S$. Let $x_1, \ldots, x_n$ be a minimal system of generators of $\mathfrak{m}_S$ and $f_j = \sum_{i=1}^{n} a_{ij}x_i$, $j = 1, \ldots, r$, a minimal system of generators of $\mathfrak{a}$. We have to show: $b_2 = \binom{n}{2} + r$. But the regular syzygies $-\bar{x}_j e_i + \bar{x}_i e_j$, $1 \leq i < j \leq n$, together with the syzygies $\sum_{i=1}^{n} \bar{a}_{ij}e_i$, $j = 1, \ldots, r$, are a minimal system of generators of all the syzygies of the residue classes $\bar{x}_1, \ldots, \bar{x}_n \in R$. Note: $-x_j e_i + x_i e_j$, $1 \leq i < j \leq n$, is a minimal system of generators of $\text{Syz}_S(x_1, \ldots, x_n)$ since $x_1, \ldots, x_n$ is a regular sequence in S.)

Conclusion. By definition, R is a complete intersection if and only if the ideal $\mathfrak{a}$ is generated by a regular sequence in S, i.e. if and only if the minimal number of generators of $\mathfrak{a}$ is equal to ht $\mathfrak{a}$. By b), this condition is equivalent with dev $R = 0$. This shows that the property of being a complete intersection is independent of the representation $R = S/\mathfrak{a}$.

The condition $\dim R + b_2 - \binom{b_1+1}{2} = 0$ can be used to define a complete intersection in the absolute sense even in the case that the local ring is not representable as a homomorphic image of a regular local ring.

6.† Let V be a module over $A[X] = A[X_1, \ldots, X_n]$. Then V has finite projective dimension, resp. finite Tor-dimension, over $A[X]$ if and only if V has finite projective dimension, resp. finite Tor-dimension, over A (with $\text{pd}_{A[X]}V \leq n + \text{pd}_A V$, resp. $\mathrm{fd}_{A[X]}V \leq n + \mathrm{fd}_A V$, where fd stands for Tor-dimension). For a canonical construction which proves this, see the beginning of §4 in [39].

Since for finite modules over a noetherian ring the projective dimension coincides with the Tor-dimension, it follows that a finitely generated flat algebra $B = A[X]/\mathfrak{A}$ over A has finite projective dimension over $A[X]$, or equivalently, that the ideal $\mathfrak{A} \subseteq A[X]$ has finite projective dimension over $A[X]$. Especially, for every finitely generated locally complete intersection B we have $\mathrm{pd}_A B < \infty$. (The last statement can be seen more directly using Koszul complexes.)

7.† In this supplement we describe a differential criterion for complete intersections going back to W.V. Vasconcelos. Let B be a finitely generated algebra over the noetherian ring A, which is of relative dimension s, i.e. $B_{\mathfrak{M}}/(A \cap \mathfrak{M})B_{\mathfrak{M}}$ has dimension s for every maximal ideal $\mathfrak{M}$ in B. Assume furthermore that B is generically smooth, i.e. that the module of Kähler differentials $\Omega_A(B)$ has rank s. Then the following conditions are equivalent: (1) B is locally a complete intersection. (2) B is A-flat and $\mathrm{pd}_B \Omega_A(B) \leq 1$. (3) B is non-degenerated over A, $\mathrm{pd}_A B < \infty$ and $\mathrm{pd}_B \Omega_A(B) \leq 1$.

Proof. Let $B = A[X]/\mathfrak{A}$. For the implication (1) $\Rightarrow$ (2) one considers the canonical exact sequence

$$\mathfrak{A}/\mathfrak{A}^2 \xrightarrow{\ j\ } B \otimes_{A[X]} \Omega_A(A[X]) \longrightarrow \Omega_A(B) \longrightarrow 0\,.$$

Since $\mathfrak{A}/\mathfrak{A}^2$ is a projective B-module of rank $n - s$ it is enough to prove that j is injective. But this is clear after tensoring the sequence with the total quotient ring L of B because $L \otimes_B \Omega_A(B)$ is free of rank s over L by generic smoothness.

That (2) implies (3) follows from Suppl. 6 and from the simple fact that flat extensions are non-degenerated. (By definition, B is n o n - d e g e n e r a t e d over A, if $\dim B_{\mathfrak{P}} = \dim A_{\mathfrak{p}} + \dim B_{\mathfrak{P}}/\mathfrak{p}B_{\mathfrak{P}}$ for every $\mathfrak{P}$ in $\mathrm{Spec}\,B$, $\mathfrak{p} := A \cap \mathfrak{P}$.)

To prove (3) $\Rightarrow$ (1) consider an ideal $\mathfrak{M} \in \mathrm{Spm}\,B \subseteq \mathrm{V}(\mathfrak{A}) \subseteq \mathrm{Spec}\,A[X]$. Because the projective dimension $\mathrm{pd}_B \Omega_A(B)$ is ≤ 1 one has $\mathfrak{A}_{\mathfrak{M}}/\mathfrak{A}_{\mathfrak{M}}^2 = W \oplus V$ where W is a free $B_{\mathfrak{M}}$-module of rank $n - s$. By a slight generalization of 6.4 (note $\mathrm{pd}_{A[X]}B < \infty$) there is a regular sequence of length $n - s$ in $\mathfrak{A}_{\mathfrak{M}}$ which generates W, but even a regular sequence of length $n-s+1$ if $V \neq 0$. The second case does not occur since B is non-degenerated over A of relative dimension s.

In the situation of this supplement we have for a locally complete intersection $B = A[X]/\mathfrak{A}$ over A an exact sequence

$$0 \longrightarrow \mathfrak{A}/\mathfrak{A}^2 \longrightarrow B^n \longrightarrow \Omega_A(B) \longrightarrow 0$$

from which the formula

$$c_A(B) = -[\Omega_A(B)]^*$$

in $\tilde{K}_0(B)$ is derived where * denotes the involution in the projective class group induced by taking duals of finite projective modules.

8.† Let B be a locally complete intersection of relative dimension s over A with the representation $B = A[X_1, \ldots, X_n]/\mathfrak{A}$. Then

$$\det c_A(B) \cong \mathrm{Ext}_{A[X]}^{n-s}(B, A[X])\,.$$

(Let $r := n - s$. The result follows from the fact that there is a canonical perfect duality

$$\bigwedge^r \mathfrak{A}/\mathfrak{A}^2 \times \mathrm{Ext}_{A[X]}^r(B, A[X]) \longrightarrow B$$

over B. This is constructed locally with respect to $\mathrm{Spec}\,B$, for which one may assume that $\mathfrak{A}/\mathfrak{A}^2$ is a free B-module of rank r generated by the residue class of a regular sequence $F_1, \ldots, F_r$ in $\mathfrak{A}$. Using the Koszul resolution of B over $A[X]$, a B-basis of the Ext-module will be represented by the residue class of the dual $e_1^* \wedge \cdots \wedge e_r^*$ of the basis $e_1 \wedge \cdots \wedge e_r \in \bigwedge^r (A[X]^r)$. Then the duality is described by the equation $\langle \bar{F}_1 \wedge \cdots \wedge \bar{F}_r, \overline{e_1^* \wedge \cdots \wedge e_r^*} \rangle = 1$. This duality is independent of the choice of the B-basis $\bar{F}_1, \ldots, \bar{F}_r$ of $\mathfrak{A}/\mathfrak{A}^2$.)

The module
$$\mathrm{Ext}_{A[X]}^{n-s}(B, A[X])$$
is the so-called relative dualizing module $\omega_{B|A}$ of B over A. If B is finite over A, then it coincides with $\mathrm{Hom}_A(B, A) = \mathrm{Ext}_A^0(B, A)$, cf. Sect. 14, Suppl. 6.

7 Graded Complete Intersections

In this section we discuss peculiarities of regular sequences and complete intersections in the case of (positively) graded A-algebras $B = \bigoplus_{m \geq 0} B_m$ with $B_0 = A$, that is, A-algebras of type

$$B = P/\mathfrak{A}, \quad P := A[X_1, \ldots, X_n],$$

where the ideal $\mathfrak{A}$ is homogeneous and contained in the irrelevant ideal $\mathfrak{M} = (X_1, \ldots, X_n)$ of P and where the X_i are homogeneous indeterminates of positive degrees γ_i, $i = 1, \ldots, n$.

As in Section 6, the base ring A is assumed to be noetherian.

First of all, let us note that in the graded case there is no difference between regular and strongly regular sequences:

7.1 Lemma *A regular sequence $f_1, \ldots, f_r$ of homogeneous elements of positive degrees $\delta_1, \ldots, \delta_r$ in a graded noetherian ring $R = \bigoplus_{m \geq 0} R_m \neq 0$ is a strongly regular sequence.*

Proof. Let $0 \leq i < r$. We must show that f_{i+1} does not belong to any associated prime ideal $\mathfrak{p}$ of $(f_1, \ldots, f_i)$. However, $\mathfrak{p}$ is homogeneous. Therefore $\mathfrak{p} + (f_{i+1}, \ldots, f_r)$ is not the unit ideal. Hence there is a (homogeneous) maximal ideal $\mathfrak{m}$ containing both $(f_1, \ldots, f_r)$ and $\mathfrak{p}$. Because $f_1, \ldots, f_r$ is a regular sequence in $R_{\mathfrak{m}}$, $f_{i+1} \notin \mathfrak{p}$. □

Furthermore, in the graded case there is no difference between locally and globally defining functions:

7.2 Theorem *Let $B = P/\mathfrak{A}$ be a graded complete intersection of relative dimension s over A with locally defining homogeneous functions $F_1, \ldots, F_{n-s}$ contained in $\mathfrak{A}$ of positive degrees $\delta_1, \ldots, \delta_{n-s}$. Then:*

(1) $F_1, \ldots, F_{n-s}$ are globally defining functions, i. e. they generate $\mathfrak{A}$. In particular, $F_1, \ldots, F_{n-s}$ is a (strongly) regular sequence.

(2) The homogeneous components B_m of B are finite stably free A-modules and in particular finite projective A-modules with rank. The corresponding Poincaré series is

$$\sum_{m \geq 0} (\mathrm{rank}_A B_m) Z^m = \frac{(1 - Z^{\delta_1}) \cdots (1 - Z^{\delta_{n-s}})}{(1 - Z^{\gamma_1}) \cdots (1 - Z^{\gamma_n})}.$$

Proof. For all statements except that B_m is stably free we may assume that A is local with maximal ideal $\mathfrak{m}$ and residue field k. To prove (1) let

$B' := P/(F_1, \ldots, F_{n-s})$. For every $m \in \mathbb{N}$ we have the canonical surjective homomorphism $B'_m \to B_m$, where B_m is finite and flat and therefore projective. Therefore $B'_m \to B_m$ is injective if this holds modulo $\mathfrak{m}$. But in $k[X_1, \ldots, X_n]$ we derive from $(F_1, \ldots, F_{n-s})_{\mathfrak{M}} = \mathfrak{A}_{\mathfrak{M}}$ the equality $(F_1, \ldots, F_{n-s}) = \mathfrak{A}$ because both ideals have homogeneous primary decompositions. This means $B'_m \equiv B_m$ modulo $\mathfrak{m}$ for all $m \in \mathbb{N}$.

That every B_m is stably free follows from the fact that the exact sequences

$$0 \longrightarrow (P/(F_1, \ldots, F_i))_{m-\delta_{i+1}} \xrightarrow{F_{i+1}} (P/(F_1, \ldots, F_i))_m$$
$$\longrightarrow (P/(F_1, \ldots, F_{i+1}))_m \longrightarrow 0$$

split, by an induction argument. The same exact sequences provide the formula for the Poincaré series, cf. Supplement 1. □

7.3 Theorem *Let* $F_1, \ldots, F_n \in P = A[X_1, \ldots, X_n]$ *be homogeneous polynomials of positive degrees* $\delta_1, \ldots, \delta_n$ *and* $B := A[X_1, \ldots, X_n]/(F_1, \ldots, F_n)$. *The following assertions are equivalent:*

(1) *B is a complete intersection over A of relative dimension 0.*
(2) *B is flat over A and $F_1, \ldots, F_n$ is a regular sequence in P.*
(3) *B is finite over A.*
(4) *$B/\mathfrak{m}B$ is finite over $A/\mathfrak{m}$ for every maximal ideal $\mathfrak{m}$ of A.*

If these conditions are fulfilled, B is a finite projective A-algebra of rank

$$\operatorname{rank}_A B = \delta_1 \cdots \delta_n / \gamma_1 \cdots \gamma_n \,.$$

Proof. By 6.13 and 7.2 only the implications (1) $\Rightarrow$ (3) and (4) $\Rightarrow$ (3) are not obvious. (1) implies (3): By 7.2, for any residue field k of A the Poincaré series of B over A coincides with the Poincaré series of $k \otimes_A B$ over k, which however is a polynomial because $k \otimes_A B$ is finite over k. (4) implies (3): The Poincaré series of $B/\mathfrak{m}B$ over $A/\mathfrak{m}$ are the same for all $\mathfrak{m} \in \operatorname{Spm} A$. Therefore there is an m_0 with $(B/\mathfrak{m}B)_m = B_m/\mathfrak{m}B_m = 0$ for $m \geq m_0$ and $\mathfrak{m} \in \operatorname{Spm} A$, which however means $B_m = 0$ for $m \geq m_0$.

The additional remark about $\operatorname{rank}_A B$ results from 7.2(2) by evaluating the Poincaré series at $Z = 1$. □

7.4 Proposition *Let* $F_1, \ldots, F_{n-s} \in A[X_1, \ldots, X_n]$ *be a regular sequence of homogeneous polynomials of positive degrees and*

$$B := A[X]/(F_1, \ldots, F_{n-s}) \,.$$

If $Q(A)$ *denotes the total ring of fractions of A, then*

$$Q(A) \otimes_A B = Q(A)[X]/(F_1, \ldots, F_{n-s})$$

is a complete intersection of relative dimension s over $Q(A)$ (with globally defining functions $F_1, \ldots, F_{n-s}$).

Proof. We may assume right away that $A = Q(A)$. Then we have only to show that B is flat over A. But B has finite projective dimension over $A[X]$, because $F_1, \ldots, F_{n-s}$ is a (strongly) regular sequence. Hence B and all its homogeneous components B_m, which are finite over A, have finite projective dimension over A. The assumption on A implies therefore, that all the B_m, $m \geq 0$, and B are even A-projective. □

Any graded complete intersection can be viewed as a specialization of a complete intersection which is a homogeneous extension of polynomial algebras:

7.5 Theorem *Let A be a noetherian ring and let*

$$B = A[X_1, \ldots, X_n]/(F_1, \ldots, F_{n-s})$$

be a graded complete intersection over A of relative dimension s with homogeneous polynomials $F_1, \ldots, F_{n-s}$ of positive degrees $\delta_1, \ldots, \delta_{n-s}$. Then the homogeneous A-algebra homomorphism

$$A[Y_1, \ldots, Y_{n-s}] \longrightarrow A[X_1, \ldots, X_n]$$

with $Y_i \mapsto F_i$, $\deg Y_i := \delta_i$, is a complete intersection of relative dimension s with canonical representation

$$A[X_1, \ldots, X_n] = A[Y_1, \ldots, Y_{n-s}][X_1, \ldots, X_n]/(Y_1 - F_1, \ldots, Y_{n-s} - F_{n-s}).$$

Especially, $A[Y_1, \ldots, Y_{n-s}] \to A[X_1, \ldots, X_n]$ is a flat extension.

Proof. By 6.2(2) we only have to prove that $A[Y] \to A[X]$ is flat. By Bourbaki's criterion, namely Theorem 1 in Ch. III, § 5, no. 2 of [4], this is equivalent to the vanishing of $\operatorname{Tor}_1^{A[Y]}(A, A[X])$ since $A[X]/(Y)A[X] = B$ is flat over $A = A[Y]/(Y)$ and $A[X]$ is (Y)-adically separable. The Tor-group may be computed as the first homology group of the Koszul complex

$$A[X] \otimes_{A[Y]} K(A[Y]; Y_1, \ldots, Y_{n-s}) = K(A[X]; F_1, \ldots, F_{n-s})$$

which is acyclic, see further below. □

We remark that 7.5 does not extend to the inhomogeneous case as is shown by the simple example $F_1 := X_1 X_2, F_2 := X_2 - 1$ in $k[X_1, X_2]$, k any field.

7.6 Corollary *If $B = A[X_1, \ldots, X_n]/(F_1, \ldots, F_{n-s})$ is a graded complete intersection over A of relative dimension s with homogeneous polynomials $F_1, \ldots, F_{n-s}$ of positive degrees then $A[X_1, \ldots, X_n]/(F_1, \ldots, F_{n-t})$ is a graded complete intersection over A of relative dimension t for all t with $s \leq t \leq n$.*

Proof. 7.6 follows from 7.5 since with $A[Y_1, \ldots, Y_{n-s}] \to A[X]$ the restriction $A[Y_1, \ldots, Y_{n-t}] \to A[X]$ is a complete intersection, too. □

Like Theorem 7.5 the corollary doesn't extend to the inhomogeneous case as for instance is shown by the example $F_1 := X_1 X_3$, $F_2 := X_2 X_3$, $F_3 := X_3 - 1$ in $k[X_1, X_2, X_3]$, k any field.

7.7 Corollary *Let $F_1, \ldots, F_{n-s} \in A[X] = A[X_1, \ldots, X_n]$ be homogeneous polynomials of positive degrees. For $i = 1, \ldots, s$ let N_i be a set of n-tuples of nonnegative integers such that the monomials X^ν, $\nu \in N_i$, have a fixed positive degree ε_i and generate an ideal containing some power of $(X_1, \ldots, X_n)$. Define*

$$G_i := \sum_{\nu \in N_i} U_{i\nu} X^\nu,$$

$i = 1, \ldots, s$, with independent indeterminates $U_{i\nu}$. Then the A-algebra $B = A[X]/(F_1, \ldots, F_{n-s})$ is a complete intersection of relative dimension s if and only if the $A(U)$-algebra $C := A(U)[X]/(F_1, \ldots, F_{n-s}, G_1, \ldots, G_s)$ over the Kronecker extension $A(U)$ of A is finite.

Proof. By 7.6 we have only to show that C is finite over $A(U)$. For this we may assume that $s > 0$ and $A = k$ is a field. Then it is enough to prove that $G = G_1$ is a non-zero-divisor modulo $F_1, \ldots, F_{n-s}$ in $k(U)[X]$, i.e. G does not belong to any prime ideal of $(F_1, \ldots, F_{n-s})$ in $k(U)[X]$. However, such an ideal is the extension of an associated prime ideal $\mathfrak{P}$ of $(F_1, \ldots, F_{n-s})$ in $k[X]$. $\mathfrak{P}$ does not contain any power of $(X_1, \ldots, X_n)$ and hence not all of the X^ν, $\nu \in N_1$. But $G \in \mathfrak{P}k(U)[X]$ implies $X^\nu \in \mathfrak{P}$ for all $\nu \in N_1$ by Supplement 1 of Section 1.   □

The proof shows that 7.7 can be modified in many ways. For example, the $U_{1\nu}$, $\nu \in N_1$, in G_1 can be replaced by different powers of a single indeterminate.

7.8 Corollary *Let $F_1, \ldots, F_{n-s} \in A[X_1, \ldots, X_n]$ be homogeneous polynomials of positive degrees. Then $B = A[X_1, \ldots, X_n]/(F_1, \ldots, F_{n-s})$ is a complete intersection of relative dimension s over A if and only if this holds modulo the maximal ideals $\mathfrak{m} \in \operatorname{Spm} A$.*

Proof. Assume that the (point) criterion for the maximal ideals is fulfilled. Then, by 7.7, 7.3 and 1.2, it follows that C is finite over $A(U)$. Now 7.7 yields that B is a complete intersection over A of relative dimension s.   □

At last we come to Koszul complexes which have to be used when studying complete intersections in detail. Let $f_1, \ldots, f_r$ be arbitrary elements of a noetherian ring R. *The Koszul complex*

$$K(f_1, \ldots, f_r) = K(R; f_1, \ldots, f_r)$$

associated to the sequence $f_1, \ldots, f_r$ is acyclic if and only if $f_1, \ldots, f_r$ is a regular sequence in R, as is well known. In this case the Koszul complex provides a finite resolution of $R/(f_1, \ldots, f_r)$ by finite free R-modules.

Assume $R = \bigoplus_{m \geq 0} R_m$ to be graded and $f_1, \ldots, f_r \in R$ to be homogeneous of degree $\delta_1, \ldots, \delta_r$. Then the Koszul complex is also graded in such a way that its derivation is homogeneous of degree 0. If $f_1, \ldots, f_r$ is a regular sequence, the homogeneous parts of the Koszul complex in degree m yield a finite resolution of $(R/(f_1, \ldots, f_r))_m$ by finite free modules over the noetherian ring R_0.

Let $B = A[X_1, \ldots, X_n]/(F_1, \ldots, F_{n-s})$ be A-flat. Then B is a complete intersection over A of relative dimension s with globally defining functions $F_1, \ldots, F_{n-s}$ if and only if the Koszul complex associated to $F_1, \ldots, F_{n-s}$ is acyclic. In the graded case as described in 7.2 we get:

7.9 Proposition *Let* $B = A[X_1, \ldots, X_n]/(F_1, \ldots, F_{n-s}) = \bigoplus_{m \geq 0} B_m$ *be a graded complete intersection of relative dimension s where $F_1, \ldots, F_{n-s}$ are homogeneous of positive degree. Then the m-th homogeneous part of the Koszul complex* $K(F_1, \ldots, F_{n-s})$ *is a finite free A-resolution of B_m.*

This proves again that every B_m is a stably free A-module.

Supplements

1. a) Let $R = \bigoplus_{m \geq 0} R_m$ be a noetherian graded A-algebra with $A = R_0$, $V = \bigoplus_{m \in \mathbb{Z}} V_m$ a finite graded R-module and $f_1, \ldots, f_r$ homogeneous elements in R of positive degrees $\beta_1, \ldots, \beta_r$ which form a regular sequence on V (and hence a strongly regular sequence on V). All the (finite) A-modules V_m have a rank over A if and only if all the homogeneous parts $\bar{V}_m$ of
$$\bar{V} := V/(f_1, \ldots, f_r)V = \bigoplus_{m \in \mathbb{Z}} \bar{V}_m$$
have a rank over A. If this is the case the following formula holds:
$$\sum_{m \in \mathbb{Z}} (\text{rank}_A \bar{V}_m) Z^m = (1 - Z^{\beta_1}) \cdots (1 - Z^{\beta_r}) \cdot \sum_{m \in \mathbb{Z}} (\text{rank}_A V_m) Z^m.$$
(It suffices to consider the case $r = 1$, in which case one uses the exact sequences $0 \to V_{m-\beta_1} \xrightarrow{f_1} V_m \to \bar{V}_m \to 0$.)

b) From a) one deduces the formula for the Poincaré series in Theorem 7.2(2) because the graded polynomial algebra $A[X_1, \ldots, X_n]$ has the Poincaré series $1/(1 - Z^{\gamma_1}) \cdots (1 - Z^{\gamma_n})$, $\gamma_i = \deg X_i > 0$. (The last formula can also be proved by a) or simply by counting monomials.)

2. Let $B = A[X_1, \ldots, X_n]/(F_1, \ldots, F_n)$ be a graded complete intersection of relative dimension 0 and $\gamma_i := \deg X_i > 0$, $\delta_j := \deg F_j > 0$. Then the Poincaré series
$$\mathcal{P} = \frac{(1 - Z^{\delta_1}) \cdots (1 - Z^{\delta_n})}{(1 - Z^{\gamma_1}) \cdots (1 - Z^{\gamma_n})}$$
is a polynomial of degree $\sigma = (\delta_1 + \cdots + \delta_n) - (\gamma_1 + \cdots + \gamma_n)$, see 7.3. (This can be seen directly in the following way: For $d \in \mathbb{N}^*$ let $\Phi_d \in \mathbb{Z}[Z]$ be the d-th cyclotomic polynomial. Then
$$Z^m - 1 = \prod_{d | m} \Phi_d.$$

Therefore, one has to prove that for every $d \in \mathbb{N}^*$ the number r of indices i with $d|\gamma_i$ is less than or equal to the number s of indices j with $d|\delta_j$. Assume $r > s$ and $d|\gamma_1, \ldots, d|\gamma_r$, $d|\delta_1, \ldots, d|\delta_s$. Then every monomial of F_j, $j > s$, contains one of the indeterminates $X_{r+1}, \ldots, X_n$, that is, $F_{s+1}, \ldots, F_n \in (X_{r+1}, \ldots, X_n)$, contradiction.)

The polynomial $\mathcal{P}$ is self-reciprocal in its coefficients. $(Z^\sigma \mathcal{P}(1/Z) = \mathcal{P}(Z).)$

3.† Let $B = \bigoplus_{m \geq 0} B_m$ be a noetherian graded algebra over $A = B_0$. Besides the projective class group $\tilde{K}_0(B)$ one has to consider the group

$$\tilde{K}_0^{\mathrm{gr}}(B)$$

of classes of graded finite projective B-modules. Two such modules P, Q define the same class in $\tilde{K}_0^{\mathrm{gr}}(B)$, if there are graded finite free B-modules V, W such that there is a homogeneous isomorphism $P \oplus V \cong Q \oplus W$ of degree zero. (V and W are required to have free bases consisting of homogeneous elements of arbitrary degrees.) A finite graded projective B-module P represents zero in $\tilde{K}_0^{\mathrm{gr}}(B)$, if there is a graded free B-module V, say, of rank r, such that $P \oplus V$ is free (as a graded B-module). In this case, for every graded free B-module V' of rank r the module $P \oplus V'$ is free. (This follows from the fact that a finite graded projective B-module Q is free as a graded module if and only if Q/B_+Q is A-free.)

The canonical homomorphism

$$\tilde{K}_0^{\mathrm{gr}}(B) \ \overset{\pi}{\longrightarrow} \ \tilde{K}_0(A)$$

induced by the augmentation $B \to A$ with kernel B_+ is an isomorphism. (π is surjective because the inclusion $A \to B$ induces a section for π. The injectivity is again a consequence of the fact that a finite graded projective B-module P is B-free if P/B_+P is A-free.)

The composition $\tilde{K}_0^{\mathrm{gr}}(B) \to \tilde{K}_0(B) \to \tilde{K}_0(A)$ of canonical homomorphisms is the isomorphism π. In particular,

$$\tilde{K}_0^{\mathrm{gr}}(B) \ \longrightarrow \ \tilde{K}_0(B)$$

is an embedding which splits.

From now on we identify $\tilde{K}_0^{\mathrm{gr}}(B)$ with its image in $\tilde{K}_0(B)$, which is nothing else but the canonical image of $\tilde{K}_0(A) \to \tilde{K}_0(B)$ (which does not depend on the grading of B under consideration).

4.† Let B be a locally complete intersection over A. If $B = \bigoplus_{m \geq 0} B_m$ is graded with $B_0 = A$, then the canonical class $c_A(B) \in \tilde{K}_0(B)$ actually is an element of the subgroup $\tilde{K}_0^{\mathrm{gr}}(B)$, cf. Suppl. 3. (If B is represented by $A[X]/\mathfrak{A}$, $\mathfrak{A}$ homogeneous, then $c_A(B)$ is represented by the graded module $\mathrm{Hom}_B(\mathfrak{A}/\mathfrak{A}^2, B)$.) Consequences are:

(1) The canonical class $c_A(B) \in \tilde{K}_0(B)$ belongs to the image of $\tilde{K}_0(A) \to \tilde{K}_0(B)$ if there is a grading $B = \bigoplus_{m \geq 0} B_m$ with $B_0 = A$.

(2) If B is a complete intersection over A and has a grading $B = \bigoplus_{m \geq 0} B_m$ with $B_0 = A$, then there is a homogeneous system of globally defining functions for B over A. (This follows from the characterization of the zero element of $\tilde{K}_0^{\mathrm{gr}}(B)$ in Suppl. 3 and by Lemma 6.5.)

(3) If $B = \bigoplus_{m \geq 0} B_m$ with $B_0 = A$ and if $\tilde{K}_0(A) = 0$, then B is a complete intersection over A. (Remark. It is well known, that $\tilde{K}_0(A) = 0$ for rings which

are polynomial rings over a regular ring R such that $\tilde{K}_0(R) = 0$. In particular, $\tilde{K}_0(R[Y_1, \ldots, Y_m]) = 0$ for principal ideal domains, fields or, more generally, arbitrary regular local rings.)

5. Let $f_1, \ldots, f_r$ be a regular sequence of homogeneous elements of positive degrees in the graded noetherian ring $R = \bigoplus_{m \geq 0} R_m$. If the ideal $\sum_{i=1}^r R f_i$ is prime in R then all the ideals $\sum_{i=1}^j R f_i$, $j = 0, \ldots, r$, are prime in R. (One has to show: If a positively graded ring R contains a prime element p (i.e. a non-zero-divisor generating a prime ideal) which in addition is homogeneous of positive degree, then R is an integral domain. But, for homogeneous zero divisors $a \neq 0$, $b \neq 0$ with $ab = 0$ and $\deg a + \deg b$ minimal one gets $a = a'p$ or $b = b'p$, hence $a'b = 0$ or $ab' = 0$, contradiction.)

6. Let $f_1, \ldots, f_r$ be a regular sequence of homogeneous elements of positive degrees in the graded noetherian ring $R = \bigoplus_{m \geq 0} R_m$ and $G_1, \ldots, G_r$ arbitrary homogeneous polynomials without constant terms in the graded polynomial ring $P = R[X_1, \ldots, X_n]$, $\deg X_i > 0$, such that $\deg f_j = \deg G_j$, $j = 1, \ldots, r$.

a) $f_1 + G_1, \ldots, f_r + G_r$ is a regular sequence in P. ($X_1, \ldots, X_n, f_1, \ldots, f_r$ is a regular sequence in P, thus $X_1, \ldots, X_n, f_1 + G_1, \ldots, f_r + G_r$, too. Then the sequence $f_1 + G_1, \ldots, f_r + G_r$ is also regular; one has to use 7.1.)

In particular, if $F_1, \ldots, F_{n-s}$ is a regular sequence of homogeneous polynomials of positive degrees in a graded polynomial algebra $Q = A[X_1, \ldots, X_n]$ over a noetherian ring A then the same is true for $F_1 \otimes 1 - 1 \otimes F_1, \ldots, F_{n-s} \otimes 1 - 1 \otimes F_{n-s}$ in $Q \otimes_A Q$.

b) If the ideal $R f_1 + \cdots + R f_r$ is prime in R then $P(f_1 + G_1) + \cdots + P(f_r + G_r)$ is a prime ideal in P. (Suppl. 5.)

In particular, if $F_1, \ldots, F_{n-s}$ is a regular sequence of homogeneous polynomials of positive degrees in a graded polynomial algebra $Q = A[X_1, \ldots, X_n]$ over a noetherian integral domain A, which generate a prime ideal, then the elements $F_1 \otimes 1 - 1 \otimes F_1, \ldots, F_{n-s} \otimes 1 - 1 \otimes F_{n-s}$ generate a prime ideal in $Q \otimes_A Q$.

8 Generic Regular Sequences

Resultants can be defined starting from a graded polynomial algebra

$$Q[T] = Q[T_0, \ldots, T_n],$$

where $T_0, \ldots, T_n$ are indeterminates of positive degrees $\gamma_0, \ldots, \gamma_n$ which are called w e i g h t s, and g e n e r i c homogeneous polynomials $F_0, \ldots, F_n$ of positive degrees $\delta_0, \ldots, \delta_n$. That is, the F_j are polynomials

$$F_j = \sum_{\nu \in N_j} U_{j\nu} T^\nu$$

where N_j is a finite set of tuples $\nu = (\nu_0, \ldots, \nu_n) \in \mathbb{N}^{n+1} \setminus \{0\}$ such that

$$\langle \nu, \gamma \rangle := \nu_0 \gamma_0 + \cdots + \nu_n \gamma_n = \delta_j$$

and where $U_{j\nu}$ are independent indeterminates over $\mathbb{Z}$: the ground ring is the polynomial ring

$$Q = \mathbb{Z}[U_{j\nu}]_{j,\nu}.$$

To avoid trivial cases we will assume throughout that the sets N_j are non-empty, i.e. that the polynomials F_j are non-zero. Sometimes it will be useful to switch to

$$Q_{(k)} = k \otimes_{\mathbb{Z}} Q = k[U_{j\nu}]_{j,\nu}$$

where k is a noetherian (commutative) ring. To be able to refer to weights and degrees in a convenient way we set

$$\gamma := (\gamma_0, \ldots, \gamma_n), \quad \delta := (\delta_0, \ldots, \delta_n).$$

For any set γ of weights and a given degree $d \in \mathbb{N}$ let

$$N_d(\gamma) := \{\nu \in \mathbb{N}^{n+1} : \langle \nu, \gamma \rangle = d\}.$$

Then $N_j \subseteq N_{\delta_j}(\gamma)$. The set N_j will be called s a t u r a t e d, if $N_j = N_{\delta_j}(\gamma)$. A situation where all the N_j involved are saturated will be referred to also as a s a t u r a t e d s i t u a t i o n or a s a t u r a t e d c a s e.

The resultant of $F_0, \ldots, F_n$ is defined if the polynomials form a regular sequence, the N_j being saturated or not. Some of the preparatory techniques and concepts we need for the construction, e. g. a suitable characterization of regular sequences, admissibility of pairs γ, δ and the handling of the principal component, do not only work for generic homogeneous polynomials but more generally for arbitrary g e n e r i c p o l y n o m i a l s, i.e.

polynomials with indeterminate coefficients. These techniques, described already in [43], will be developed during the current and the next section. The s u p p o r t of $\nu = (\nu_0, \ldots, \nu_n) \in \mathbb{N}^{n+1}$ is defined as

$$\operatorname{supp} \nu := \{i : \nu_i \neq 0\}.$$

For an arbitrary ground ring $A \neq 0$ and a subset $I \subseteq \{0, \ldots, n\}$ the monomial $T^\nu \in A[T] = A[T_0, \ldots, T_n]$ is contained in the ideal $(T_i ; i \in I) \subseteq A[T]$ generated by the indeterminates T_i, $i \in I$, if and only if $I \cap \operatorname{supp} \nu \neq \emptyset$. This is a *combinatorial condition for the ideal membership* $T^\nu \in (T_i ; i \in I)$. In particular, it depends only on the support of the exponent ν.

More generally, for $N \subseteq \mathbb{N}^{n+1}$ and the monomial ideal $\mathfrak{I}_N$ generated by the monomials T^ν, $\nu \in N$, the inclusion $\mathfrak{I}_N \subseteq (T_i ; i \in I)$ can be characterized by the combinatorial condition: $I \cap \operatorname{supp} \nu \neq \emptyset$ for every $\nu \in N$.

For $\nu \in \mathbb{N}^{n+1}$ let ν_{red} be defined by the characteristic function

$$\nu_{\mathrm{red}} = e_{\operatorname{supp} \nu}$$

of $\operatorname{supp} \nu$. Thus, if $\nu = (\nu_0, \ldots, \nu_n)$ then $\nu_{\mathrm{red},i} = 1$ for $i \in \operatorname{supp} \nu$ and $\nu_{\mathrm{red},i} = 0$ otherwise. The corresponding monomial is

$$T^{\nu_{\mathrm{red}}} = \prod_{i \in \operatorname{supp} \nu} T_i.$$

If $N \subseteq \mathbb{N}^{n+1}$ let

$$N_{\mathrm{red}} := \{\nu_{\mathrm{red}} : \nu \in N\}.$$

For a field K the ideal $\mathfrak{I}_{N_{\mathrm{red}}} \subseteq K[T]$ has the same radical as $\mathfrak{I}_N$. (The ideal $\mathfrak{I}_{N_{\mathrm{red}}}$ coincides even with the radical of $\mathfrak{I}_N$.)

The first lemma characterizes generic regular sequences:

8.1 Lemma *Let* $F_0, \ldots, F_r$, $r \leq n$, *be polynomials in* $T_0, \ldots, T_n$ *with indeterminate coefficients of the form*

$$F_j = \sum_{\nu \in N_j} U_{j\nu} T^\nu$$

over $Q = \mathbb{Z}[U_{j\nu}]_{j,\nu}$, *where* N_j *is a non-empty finite subset of* $\mathbb{N}^{n+1} \setminus \{0\}$, $j = 0, \ldots, r$. *By* $\mathfrak{I}_j$ *we denote the ideal* $\mathfrak{I}_{N_j}$ *in* $\mathbb{Z}[T]$ *generated by the monomials* T^ν, $\nu \in N_j$. *The following conditions are equivalent:*

(1) $F_0, \ldots, F_r$ *form a regular sequence in* $Q[T]$.

(1') $F_0, \ldots, F_r$ *form a regular sequence in* $Q_{(k)}[T]$ *for every noetherian ring* $k \neq 0$.

(1'') *There is a noetherian ring* $A \neq 0$ *and a specialization* $Q \to A$ *such that* $F_0, \ldots, F_r$ *form a regular sequence in* $A[T]$.

(2) $F_0, \ldots, F_r$ *form a regular sequence in the localization* $L[T]_{(T_0, \ldots, T_n)}$ *where* L *denotes the rational function field* $K(U_{j\nu})_{j,\nu}$, K *any field.*

(3) There is a field L and a specialization $Q \to L$ such that $F_0, \ldots, F_r$ form a regular sequence in the localization $L[T]_{(T_0, \ldots, T_n)}$.

(4) There is a field L such that for every subset $J \subseteq \{0, \ldots, r\}$ the ideal $\mathfrak{I}_J L[T]$ generated by $\mathfrak{I}_J := \sum_{j \in J} \mathfrak{I}_j$ has height $\geq |J|$.

(4′) For every field K and every subset $J \subseteq \{0, \ldots, r\}$ the ideal $\mathfrak{I}_J K[T]$ has height $\geq |J|$.

(4″) For every non-empty proper subset I of $\{0, \ldots, n\}$ the set
$$\{j \ : \ 0 \leq j \leq r, \ \mathfrak{I}_j \subseteq (T_i \ : \ i \in I)\} \ =$$
$$\{j \ : \ 0 \leq j \leq r, \ I \cap \operatorname{supp} \nu \neq \emptyset \quad \text{for all} \ \ \nu \in N_j\}$$
contains $\leq |I|$ elements.

(5) $F_0, \ldots, F_r$ form a strongly regular sequence in $Q_{(K)}[T]$, K any field.

Proof. (1′) implies (2) because $F_0, \ldots, F_r$ form by (1′) a regular sequence in $L[T]$ which however is contained in the ideal $(T_0, \ldots, T_n)$.

(3) is weaker than (2). (1) is weaker than (1′), and (1) implies (3) (take for L the quotient field of Q). (1′) implies (1″).

(1″) implies (3): Flat extensions preserve regular sequences. Thus we may assume that A is a local artinian ring. $F_0, \ldots, F_r$ generate an ideal of height $r + 1$ in $A[T]$ and therefore in $L[T]$ over the residue field L of A. By Macaulay's Theorem $F_0, \ldots, F_r$ form a regular sequence in $L[T]$.

(3) implies (4): Let L be as in condition (3) and let $J \subseteq \{0, \ldots, r\}$ be given. Let $\mathfrak{P}$ be a minimal prime ideal of $\mathfrak{I}_J L[T] \supseteq \sum_{j \in J} F_j L[T]$. Then $\mathfrak{P} \subseteq (T_0, \ldots, T_n)$. Because $F_j, j \in J$, is a part of the regular sequence $F_0, \ldots, F_r$ in $L[T]_{(T_0, \ldots, T_n)}$, the extension of $\mathfrak{P}$ in $L[T]_{(T_0, \ldots, T_n)}$ and therefore $\mathfrak{P}$ itself has height $\geq |J|$.

(4) and (4′) are both equivalent to (4″), because, by simple arguments, the height of a monomial ideal in a polynomial ring over a field is $\leq s$ if and only if it is contained in an ideal generated by at most s of the indeterminates.

(4′) implies (5): By induction on $|J|$ we show that the ideal $\mathfrak{F}_J$ generated by $F_j, j \in J$, in $Q_{(K)}[T]$ has height $|J|$. This yields (5) because $Q_{(K)}[T]$ is a regular ring (and hence a Macaulay ring).

For the induction step from i to $i+1$ we may assume that $J = \{0, \ldots, i\}$. Let $\mathfrak{P}$ be a minimal prime ideal of $(F_0, \ldots, F_i)$ and assume $\operatorname{ht} \mathfrak{P} \leq i$. By induction hypothesis $\mathfrak{P}$ is a minimal prime ideal of $\mathfrak{F}_{J \setminus \{s\}}$, $s = 0, \ldots, i$. Therefore $\mathfrak{P}$ is the extension of a prime ideal $\mathfrak{P}_s \subseteq K[U_{j\nu} : j \in J \setminus \{s\}$, $\nu \in N_j][T]$. By the Lemma below $\mathfrak{P}$ is the extension of a prime ideal $\mathfrak{p} \subseteq K[T]$ of height $\leq i$. In particular, the coefficients T^μ of $F_s \in K[T][U_{j\nu}]$ belong to $\mathfrak{p}$ for $s = 0, \ldots, i$, which means that $\mathfrak{p}$ contains $\mathfrak{I}_J K[T]$. This contradicts the assumption that $\mathfrak{I}_J K[T]$ has height $\geq i + 1$.

(5) implies (1′): Let k be an arbitrary noetherian ring. By (5) for any

prime ideal $\mathfrak{p} \subseteq k$ the sequence $F_0,...,F_r$ is strongly regular in $Q_{(k_\mathfrak{p}/\mathfrak{p}k_\mathfrak{p})}[T]$. By Nagata's Lemma 6.3 it follows that it is a regular sequence in $Q_{(k_\mathfrak{p})}[T]$. Now it is clear that $F_0, \ldots, F_r$ is a regular sequence in $Q_{(k)}[T]$.   □

8.2 Lemma *Let $\mathfrak{A}$ be an ideal in the polynomial algebra $A[X] = A[X_i : i \in I]$ and $I = I_1 \uplus \cdots \uplus I_m$ a partition of I. If $\mathfrak{A}$ is an extension of an ideal $\mathfrak{A}_\mu$ of $A[X_i : i \in I \setminus I_\mu]$ for $\mu = 1, \ldots, m$, then $\mathfrak{A}$ is the extension of the ideal $\mathfrak{a} := \mathfrak{A} \cap A \subseteq A$.*

Proof by induction on m. For $m = 1$ there is nothing to prove. Now let $m = 2$ and consider some $F \in \mathfrak{A}_1$. We must show that F has coefficients in $\mathfrak{a}$. Because of $F \in \mathfrak{A} = \mathfrak{A}_2 A[X]$ there is a representation $F = \sum_\nu G_\nu X^\nu$ with $G_\nu \in \mathfrak{A}_2$ and exponents $\nu \in \mathbb{N}^{(I_2)}$. Comparing coefficients yields $G_\nu \in \mathfrak{A} \cap A = \mathfrak{a}$ and thus $F \in \mathfrak{a}A[X]$. To complete the induction be left to the reader.   □

A result related to Lemma 8.1 was proved by C.T.C. Wall in [47].

Lemma 8.1 leads to:

Definition A sequence $N_0, \ldots, N_r$, $r \leq n$, of non-empty finite subsets of $\mathbb{N}^{n+1} \setminus \{0\}$ is called a d m i s s i b l e, if it satisfies the equivalent conditions of Lemma 8.1. In this case the corresponding sequence $F_0, \ldots, F_r$ of generic polynomials is also called a d m i s s i b l e.

Admissibility is a condition on the sequence $N_0, \ldots, N_r$ of purely combinatorial character, as is expressed by (4″) in 8.1. Admissibility does not depend on the order of the elements of the sequence. $N_0, \ldots, N_r$ is admissible if and only if the sequence $N_{0,\text{red}}, \ldots, N_{r,\text{red}}$ of the sets of reduced exponents is admissible. If $N_0, \ldots, N_r$ is admissible then every sequence $N_0', \ldots, N_r'$ with $N_j \subseteq N_j'$, $j = 0, \ldots, r$, is admissible. Furthermore, $N_0, \ldots, N_r$ remains admissible if one replaces their elements ν by exponents $\tilde{\nu}$ with $\emptyset \neq \text{supp}\,\tilde{\nu} \subseteq \text{supp}\,\nu$. Every subsequence of an admissible sequence is admissible.

We can leave it to the reader to formulate the corresponding properties for admissible sequences of generic polynomials. For example, let us note that admissibility of a sequence of generic polynomials does not depend on the order of the polynomials and that subsequences remain admissible.

In the special case of homogeneous generic polynomials, a pair of sequences $\gamma = (\gamma_0, \ldots, \gamma_n)$, $\delta = (\delta_0, \ldots, \delta_r)$ of weights γ_i and degrees δ_j is called a d m i s s i b l e if the sequence $N_{\delta_0}(\gamma), \ldots, N_{\delta_r}(\gamma)$ of corresponding saturated sets is admissible. If subsets $N_0 \subseteq N_{\delta_0}, \ldots, N_r \subseteq N_{\delta_r}$ form an admissible sequence, then the pair γ, δ is necessarily admissible. To phrase it differently: Admissible sequences of generic homogeneous polynomials $F_0, \ldots, F_r$ can only exist, if their sequences of weights and degrees form an admissible pair.

For the handling of generic regular sequences the next lemma is useful.

8.3 Lemma *Let $F_0, \ldots, F_r$, $r \leq n$, be an admissible sequence of generic polynomials and let k be an arbitrary noetherian ring. Then every non-zero-divisor a in k is a non-zero-divisor in the algebra $Q_{(k)}[T]/(F_0, \ldots, F_r)$.*

Proof. Let $\mathfrak{P}$ be an associated prime ideal of $(F_0, \ldots, F_r)$ and assume $a \in \mathfrak{P}$. By 8.1(1′) the elements $F_0, \ldots, F_r$ form a regular sequence in $Q_{(k/ak)}[T]_{\mathfrak{P}}$, which contradicts the fact that $F_0, \ldots, F_r$ is a maximal regular sequence in $Q_{(k)}[T]_{\mathfrak{P}}$. □

An admissible sequence of generic polynomials $F_0, \ldots, F_r$ is a regular sequence in $Q_{(k)}[T]$ for an arbitrary noetherian ring k, cf. condition (1′) in Lemma 8.1. Since every subsequence of $F_0, \ldots, F_r$ is also admissible, 6.1 implies:

8.4 Proposition *Let $F_0, \ldots, F_r$, $r \leq n$, be an admissible sequence of generic polynomials and let k be an arbitrary noetherian ring $\neq 0$. Then $F_0, \ldots, F_r$ is a strongly regular sequence in $Q_{(k)}[T]$.*

8.5 Corollary *In the situation of Proposition 8.4 assume that k is an integral domain. Then $F_0, \ldots, F_r$ is a strongly regular sequence in $Q_{(k)}[T]$ generating an unmixed ideal of pure codimension $r + 1$.*

Proof. The corollary is clear if k is a field since every strongly regular sequence of length $r + 1$ in the polynomial ring $Q_{(k)}[T]$ over k generates an unmixed ideal of pure codimension $r + 1$. The general case will be reduced to this special case using 8.3. □

There is more to say about the Lasker-Noether decomposition of an ideal generated by a generic regular sequence. Let us consider arbitrary generic polynomials $F_0 \neq 0, \ldots, F_r \neq 0$ and a noetherian integral domain k. Among the primary components of the ideal $(F_0, \ldots, F_r)$ in $Q_{(k)}[T]$ there is a distinguished one, which we will denote by $\mathfrak{H}^{\star}$ and which is given by

$$\mathfrak{H}^{\star} := (F_0, \ldots, F_r)Q_{(k)}[T, T^{-1}] \cap Q_{(k)}[T],$$

making use of the ring $Q_{(k)}[T, T^{-1}] = Q_{(k)}[T_0, \ldots, T_n, T_0^{-1}, \ldots, T_n^{-1}] = Q_{(k)}[T]_{T_0 \cdots T_n}$ of Laurent polynomials. That $\mathfrak{H}^{\star}$ really is a primary component simply follows from the fact that $F_0, \ldots, F_r$ generate a prime ideal of height $r + 1$ in $Q_{(k)}[T, T^{-1}]$, which can be proved in the following way: In $Q_{(k)}[T, T^{-1}]$ the polynomial F_j may be replaced by

$$T^{-\nu(j)}F_j = U_{j\nu(j)} - G_j,$$

where $\nu(j)$ is an arbitrary element of $N_j (\neq \emptyset)$. Note that G_j does not contain any of the variables $U_{l\nu(l)}$, $l = 0, \ldots, r$. Let $Q^\flat_{(k)}$ denote the polynomial ring

$$Q^\flat_{(k)} := k[U_{j\nu} : j = 0, \ldots, r, \ \nu \in N_j, \nu \neq \nu(j)].$$

Then $G_j \in Q^\flat_{(k)}[T, T^{-1}]$ and $Q_{(k)}[T, T^{-1}]/(F_0, \ldots, F_r)$, which can be written in the form

$$Q^\flat_{(k)}[T, T^{-1}][U_{0\nu(0)}, \ldots, U_{r\nu(r)}]/(U_{0\nu(0)} - G_0, \ldots, U_{r\nu(r)} - G_r),$$

is canonically isomorphic to $Q^\flat_{(k)}[T, T^{-1}]$ and hence certainly an integral domain.

It follows that *the primary component $\mathfrak{H}^\star$ actually is prime*. (We assumed the ring k to be an integral domain.)

Definition Let k be an integral domain. The primary component

$$\mathfrak{H}^\star := (F_0, \ldots, F_r)Q_{(k)}[T, T^{-1}] \cap Q_{(k)}[T]$$

is called the **principal component** (Hauptkomponente) of $(F_0, \ldots, F_r)$.

$\mathfrak{H}^\star$ coincides with $(F_0, \ldots, F_r)$ if and only if $(F_0, \ldots, F_r)$ is a prime ideal. In the case $r < n$ this situation is characterized in the following lemma.

8.6 Lemma *Assume $r < n$. Let $F_0, \ldots, F_r$ be generic polynomials in $T_0, \ldots, T_n$ over $Q = \mathbb{Z}[U_{j\nu}]_{j,\nu}$ with non-empty defining sets $N_0, \ldots, N_r$ of exponents in $\mathbb{N}^{n+1} \setminus \{0\}$. By $\mathfrak{I}_j$ we denote the ideal $\mathfrak{I}_{N_j}$ in $\mathbb{Z}[T]$ generated by the monomials T^ν, $\nu \in N_j$. The following conditions are equivalent:*

(1) $F_0, \ldots, F_r$ *generate a prime ideal in $Q[T]$.*

(1') $F_0, \ldots, F_r$ *generate a prime ideal in $Q_{(k)}[T]$ for every noetherian integral domain k.*

(1'') $F_0, \ldots, F_r$ *generate a prime ideal in $Q_{(k)}[T]$ for some noetherian integral domain k.*

(2) *There is a field L such that for every non-empty subset J of $\{0, \ldots, r\}$ the ideal $\mathfrak{I}_J L[T]$ generated by $\mathfrak{I}_J := \sum_{j \in J} \mathfrak{I}_j$ has height $\geq |J| + 1$.*

(2') *For every field K and for every non-empty subset J of $\{0, \ldots, r\}$ the ideal $\mathfrak{I}_J K[T]$ generated by $\mathfrak{I}_J$ has height $\geq |J| + 1$.*

(2'') *For every non-empty proper subset I of $\{0, \ldots, n\}$ the set*
$$\{j : 0 \leq j \leq r, \ \mathfrak{I}_j \subseteq (T_i : i \in I)\} =$$
$$\{j : 0 \leq j \leq r, \ I \cap \operatorname{supp}\nu \neq \emptyset \ \text{for all} \ \nu \in N_j\}$$
contains $\leq |I| - 1$ elements.

Proof. (1'') implies (2). For L we take the quotient field of $Q_{(k)}$. In $Q_{(k)}[T]$ the ideal $(F_0, \ldots, F_r)$ is necessarily of height $r + 1$, because it remains prime in $Q_{(k)}[T, T^{-1}]$, where however the non-zero polynomials F_j behave like independent indeterminates. (This is the point where the assumption is used that all the defining sets of exponents are non-empty.) Therefore, $(F_0, \ldots, F_r)$ is a prime ideal of height $r + 1$ in $L[T]$ and $F_0, \ldots, F_r$ is a regular sequence in $L[T]$. By the implication $(1'') \Rightarrow (4)$ in 8.1, for every subset $J \subseteq \{0, \ldots, r\}$ the ideal $\mathfrak{I}_J L[T]$ has height $\geq |J|$. Let us assume that there is a J with $J \neq \emptyset$ and ht $\mathfrak{I}_J L[T] = |J|$. Then $\mathfrak{I}_J L[T]$ is contained in a prime ideal generated by $|J|$ of the indeterminates $T_0, \ldots, T_n$. [1] The same is then true in $Q_{(k)}[T]$. We may assume $J = \{0, \ldots, s\}$ and $(F_0, \ldots, F_s) \subseteq (T_0, \ldots, T_s)$.

Let $\mathfrak{P}$ be a minimal prime ideal of $(T_0, \ldots, T_s, F_{s+1}, \ldots, F_r)$. Its codimension is $\leq r + 1$, and it contains $F_0, \ldots, F_r$. By assumption $(1'')$, $\mathfrak{P} = (F_0, \ldots, F_r)$. But $T_0 \notin (F_0, \ldots, F_r) \subseteq \sum_{j,\nu} Q[T] U_{j\nu}$.

(2) and $(2')$ are both equivalent to $(2'')$, because the height of a monomial ideal in a polynomial ring over a field is $\leq s$ if and only if it is contained in an ideal generated by at most s of the indeterminates.

(2) implies $(1')$. By 8.1 and 8.3 we may assume that k is a field. Then we proceed by induction on r. By induction hypothesis we may take for granted that $F_0, \ldots, F_{r-1}$ generate a prime ideal of height r (which is the zero ideal in case $r = 0$). To prove that F_r is a prime element in the Macaulay integral domain

$$(k[U_{j\nu} : j < r][T]/(F_0, \ldots, F_{r-1}))[U_{r\nu}]_\nu$$

we have, by Lemma 8.7 below, to show that the T-monomials T^ν, $\nu \in N_r$, generate an ideal of codimension ≥ 2 modulo $F_0, \ldots, F_{r-1}$, i.e. that these monomials together with $F_0, \ldots, F_{r-1}$ generate an ideal of codimension $\geq r + 2$. Assume that $\mathfrak{P}$ is a prime ideal containing $F_0, \ldots, F_{r-1}$ and all the monomials T^ν, $\nu \in N_r$, and hence F_r, too. Then $\mathfrak{P}$ contains one of the indeterminates $T_0, \ldots, T_n$, say T_0. By assumption (2) and Lemma 8.1 the polynomials $F_0(0, T_1, \ldots, T_n), \ldots, F_r(0, T_1, \ldots, T_n)$ generate an ideal of codimension $r + 1$. Thus $\mathfrak{P}$ has codimension $\geq r + 2$.

The other implications are clear.   □

8.7 Lemma *Let A be a noetherian integral domain and $a_1, \ldots, a_m$ elements of A generating an ideal of depth ≥ 2. Then $a_1 X_1 + \cdots + a_m X_m$ is a prime element in the polynomial algebra $A[X_1, \ldots, X_m]$.*

Proof. We may assume $a_1 \neq 0$. By a well-known lemma we have to show that $a_2 X_2 + \cdots + a_m X_m$ is a non-zero-divisor in $\bar{A}[X_2, \ldots, X_m]$, $\bar{A} := A/Aa_1$, which however is clear because $\bar{A}\bar{a}_2 + \cdots + \bar{A}\bar{a}_m$ has depth ≥ 1. □

---

[1] Even more is true: The associated prime ideals of monomial ideals in $L[T]$ are generated by indeterminates, cf. for instance [34], 2.22, Satz 37 by R. Kummer.

Note that the sequence $F_0, \ldots, F_r$ in Lemma 8.6, in the case where the conditions given are fulfilled, is necessarily admissible.

Before we derive a result like Lemma 8.6 appropriate to the case $r = n$, let us first note that for admissible generic polynomials $F_0, \ldots, F_n$ and an integral domain k the ideal $(F_0, \ldots, F_n)$ in $Q_{(k)}[T]$ necessarily has the prime ideal $(T_0, \ldots, T_n)$ of codimension $n + 1$ as an associated prime ideal. Thus there is a decomposition

$$(F_0, \ldots, F_n) = \mathfrak{Q}^\star \cap \mathfrak{T}^\star$$

in $Q_{(k)}[T]$, where $\mathfrak{Q}^\star$ is the primary component belonging to $(T_0, \ldots, T_n)$ and $\mathfrak{T}^\star$ is the intersection of the other primary components, one of which is the prime ideal $\mathfrak{H}^\star$, the principal component. Then $\mathfrak{T}^\star \subseteq \mathfrak{H}^\star$, and $\mathfrak{T}^\star$ coincides with $\mathfrak{H}^\star$ if and only if $\mathfrak{T}^\star$ is prime, or if and only if there is no associated prime ideal of $(F_0, \ldots, F_n)$ different from $(T_0, \ldots, T_n)$ and containing one of the indeterminates $T_0, \ldots, T_n$. This situation can be characterized in a way fitting Lemma 8.6:

8.8 Lemma *Assume $r = n$. Let $F_0, \ldots, F_n$ be admissible generic polynomials in $T_0, \ldots, T_n$ over $Q = \mathbb{Z}[U_{j\nu}]_{j,\nu}$ with defining sets $N_0, \ldots, N_n$ of exponents. By $\mathfrak{I}_j$ we denote the ideal $\mathfrak{I}_{N_j}$ in $\mathbb{Z}[T]$ generated by the monomials T^ν, $\nu \in N_j$. The following conditions are equivalent:*

(1) $\mathfrak{T}^\star$ *is a prime ideal in $Q[T]$.*

(1') $\mathfrak{T}^\star$ *is a prime ideal in $Q_{(k)}[T]$ for every noetherian integral domain k.*

(1'') $\mathfrak{T}^\star$ *is a prime ideal in $Q_{(k)}[T]$ for some noetherian integral domain k.*

(2) *There is a field L such that for every non-empty proper subset J of $\{0, \ldots, n\}$ the monomial ideal $\mathfrak{I}_J L[T]$ has height $\geq |J| + 1$.*

(2') *For every field K and for every non-empty proper subset J of $\{0, \ldots, n\}$ the monomial ideal $\mathfrak{I}_J K[T]$ has height $\geq |J| + 1$.*

(2'') *For every non-empty proper subset I of $\{0, \ldots, n\}$ the set*
$$\{j : 0 \leq j \leq n, \ \mathfrak{I}_j \subseteq (T_i : i \in I)\} =$$
$$\{j : 0 \leq j \leq n, \ I \cap \operatorname{supp} \nu \neq \emptyset \ \text{for all} \ \nu \in N_j\}$$
contains $\leq |I| - 1$ elements.

Proof. $(1'') \Rightarrow (2)$: For L take the quotient field of $Q_{(k)}$. By Lemma 8.1, the ideal $\mathfrak{I}_J L[T]$ has height $\geq |J|$. Assume then that $\operatorname{ht} \mathfrak{I}_J L[T] = |J|$. As in the proof of 8.6 we may assume $(F_0, \ldots, F_s) \subseteq (T_0, \ldots, T_s)$ for some s with $0 \leq s < n$. Again by Lemma 8.1, $F_j(0, \ldots, 0, T_{s+1}, \ldots, T_n)$, $j = s+1, \ldots, n$, form a regular sequence in $Q_{(k)}[T_{s+1}, \ldots, T_n] = Q_{(k)}[T]/(T_0, \ldots, T_s)$, the principal component of which defines an associated prime ideal of the ideal $(T_0, \ldots, T_s, F_{s+1}, \ldots, F_n)$ in $Q_{(k)}[T]$ of height $n + 1$ and hence an associated prime ideal of $(F_0, \ldots, F_n)$ containing $(T_0, \ldots, T_s)$ and different from $(T_0, \ldots, T_n)$, a contradiction.

(2) and (2′) are both equivalent to (2″) by the same argument used to prove the equivalence of (2), (2′) and (2″) in 8.6.

(2) $\Rightarrow$ (1′): By Lemma 8.6, every subsequence of $F_0, \ldots, F_n$ of length $m \le n$ generates a prime ideal of height m. Let $\mathfrak{P}$ be an associated prime ideal of $(F_0, \ldots, F_n)$ containing one of the indeterminates $T_0, \ldots, T_n$, say T_n. By 8.5, $\mathfrak{P}$ has height $n+1$. We have to show that $\mathfrak{P}$ is nothing else but $(T_0, \ldots, T_n)$. For this it is enough to show that all the monomials T^ν, $\nu \in N_j$, belong to $\mathfrak{P}$, $j = 0, \ldots, n$. The polynomials F_i, $i \ne j$, generate a prime ideal of height n which does not contain T_n. Therefore $\mathfrak{P}$ is a minimal prime of the ideal generated by the F_i, $i \ne j$, and by T_n. Hence $\mathfrak{P}$ is extended from a prime ideal $\mathfrak{P}' \subseteq Q'_{(k)}[T]$, $Q'_{(k)} := k[U_{i\nu} : i \ne j]$, and the coefficients T^ν, $\nu \in N_j$, of $F_j \in \mathfrak{P}$ belong to $\mathfrak{P}' \subseteq \mathfrak{P}$.

The other implications are clear. $\square$

The following definition is motivated by 8.6 and 8.8.

Definition A sequence $N_0, \ldots, N_r$, $r \le n$, of non-empty finite subsets of $\mathbb{N}^{n+1} \smallsetminus \{0\}$ is called **strictly admissible**, if the following condition is fulfilled:

For every non-empty proper subset I of $\{0, \ldots, n\}$ the set

$$\{j \,:\, 0 \le j \le r, \ I \cap \operatorname{supp} \nu \ne \emptyset \ \text{for all} \ \nu \in N_j\}$$

contains $\le |I| - 1$ elements.

In this case the corresponding sequence $F_0, \ldots, F_r$ of generic polynomials is also called **strictly admissible**.

A sequence $N_0, \ldots, N_n$ with $n \ge 1$ is strictly admissible if and only if every subsequence of length n is strictly admissible.

To strictly admissible sequences analogous remarks apply as were specified after the definition of admissible sequences above. In particular, strict admissibility does not depend on the order of sequences, is passed on to subsequences and is stable under the process of enlarging sets of exponents.

In the special case of homogeneous generic polynomials, a pair of sequences $\gamma = (\gamma_0, \ldots, \gamma_n)$, $\delta = (\delta_0, \ldots, \delta_r)$ of weights γ_i and degrees δ_j is called **strictly admissible** if the sequence $N_{\delta_0}(\gamma), \ldots, N_{\delta_r}(\gamma)$ of corresponding saturated sets of exponents is strictly admissible.

Strictly admissible sequences $F_0, \ldots, F_r$ of homogeneous generic polynomials can only exist if their sequences of weights and degrees are strictly admissible. In the classical case $\gamma = (1, \ldots, 1)$ all sequences $F_0, \ldots, F_r$, $r \le n$, of saturated homogeneous generic polynomials of positive degrees are strictly admissible.

Naturally, every strictly admissible sequence is admissible. Furthermore, if $r < n$, there is a natural link between the concepts of admissibility and strict admissibility, which stems directly from the definitions:

8.9 Lemma *Assume $r < n$. For generic polynomials $F_0 \neq 0, \ldots, F_r \neq 0$ in $T_0, \ldots, T_n$ the following conditions are equivalent:*
(1) *The sequence $F_0, \ldots, F_r$ is strictly admissible.*
(1') *The sequence $F_0, \ldots, F_r, UT_0 \cdots T_n$ is admissible.*
(1'') *The sequence $F_0, \ldots, F_r, U_iT_i$ is admissible for every $i = 0, \ldots, n$.*

This lemma also simplifies the handling of strictly admissible sequences of generic polynomials.

At the end of this section we will treat sequences of strictly admissible generic binomials in greater detail with respect to combinatorics, in order to explore this feature of regular sequences. The results will not be used in the later sections.

A g e n e r i c   b i n o m i a l is a polynomial of the type

$$F = UT^\mu + VT^\nu$$

with different exponents $\mu, \nu \in \mathbb{N}^{n+1} \setminus \{0\}$ and indeterminate coefficients U, V. The binomial F forms a strictly admissible sequence (of length 1) if and only if the supports

$$C = \operatorname{supp} \mu, \quad D = \operatorname{supp} \nu$$

are disjoint. Therefore *we will consider here only generic binomials with the additional property $C \cap D = \emptyset$.* For a subset I of

$$\Omega = \{0, \ldots, n\}$$

the binomial F belongs to the ideal generated by T_i, $i \in I$, if and only if $C \cap I \neq \emptyset$ and $D \cap I \neq \emptyset$. If this is the case, we say that the pair (C, D) r e s t s   o n I.

Then we restrict our attention to the case of a sequence of generic binomials in $T_0, \ldots, T_n$ of length n, which will be denoted in a suitable way, starting not at 0 but at 1, by

$$F_j = U_jT^{\mu(j)} + V_jT^{\nu(j)}, \quad j = 1, \ldots, n.$$

These are the generic polynomials corresponding to the sets

$$N_j = \{\mu(j), \nu(j)\}, \quad j = 1, \ldots, n$$

of exponents. The sets $N_{j,\text{red}}$ of the corresponding reduced exponents are described by the pairs

$$M_j := (C_j, D_j), \quad C_j := \operatorname{supp} \mu(j), \quad D_j := \operatorname{supp} \nu(j).$$

The sequence $F_1, \ldots, F_n$ is strictly admissible if and only if for every non-empty proper subset $I \subset \Omega$ there are at most $|I| - 1$ pairs M_j resting on the set I.

Strict admissibility of sequences of binomials can be characterized with suitable partitions of Ω. A partition $\mathfrak{Z}$ of Ω is a set of pairwise disjoint non-empty subsets covering Ω. The set of partitions of Ω is ordered in a natural way: We say, $\mathfrak{Z}_1$ *refines* $\mathfrak{Z}_2$, and write

$$\mathfrak{Z}_1 \lhd \mathfrak{Z}_2$$

for short, if every $X \in \mathfrak{Z}_1$ is contained in some $Y \in \mathfrak{Z}_2$. This means that every set of $\mathfrak{Z}_2$ is the union of some sets of $\mathfrak{Z}_1$. For a partition $\mathfrak{Z}$ of Ω and a subset $I \subseteq \Omega$ let

$$\mathfrak{Z}(I) := \{X \in \mathfrak{Z} : X \cap I \neq \emptyset\}.$$

To the sequence $M_1, \ldots, M_n$ as above there is a corresponding chain

$$\mathfrak{Z}_0 \lhd \mathfrak{Z}_1 \lhd \cdots \lhd \mathfrak{Z}_n$$

of partitions constructed recursively in the following way: $\mathfrak{Z}_0$ is simply the set $\{\{0\}, \{1\}, \ldots, \{n\}\}$, and $\mathfrak{Z}_j$ is obtained from $\mathfrak{Z}_{j-1}$ by replacing the sets of $\mathfrak{Z}_{j-1}(C_j \cup D_j)$ by their union, $j = 1, \ldots, n$.

Following a terminology introduced by Ch. Delorme in [11], we will say that the sequence $M_1, \ldots, M_n$ is **distinguished**, if the corresponding chain $\mathfrak{Z}_0, \ldots, \mathfrak{Z}_n$ fulfills the condition

$$\mathfrak{Z}_{j-1}(C_j) \cap \mathfrak{Z}_{j-1}(D_j) = \emptyset$$

for $j = 1, \ldots, n$. This is equivalent to the following condition: C_j is contained in one element of $\mathfrak{Z}_{j-1}$, and D_j is contained in another element of $\mathfrak{Z}_{j-1}$; $j = 1, \ldots, n$. In this case the partition $\mathfrak{Z}_j$ contains exactly $n + 1 - j$ elements; in particular, $\mathfrak{Z}_n = \{\Omega\}$.

Consider $M_1 := (\{0\}, \{1\})$, $M_2 := (\{0, 1\}, \{3\})$ and $M_3 := (\{0, 3\}, \{2\})$ as an example. To M_1, M_2, M_3 correspond $\mathfrak{Z}_0 = \{\{0\}, \{1\}, \{2\}, \{3\}\}$, $\mathfrak{Z}_1 = \{\{0, 1\}, \{2\}, \{3\}\}$, $\mathfrak{Z}_2 = \{\{0, 1, 3\}, \{2\}\}$, $\mathfrak{Z}_3 = \{\Omega\}$, hence M_1, M_2, M_3 is distinguished. But to M_1, M_3, M_2 corresponds $\mathfrak{Z}_0, \mathfrak{Z}_1, \mathfrak{Z}_2', \mathfrak{Z}_3$, where $\mathfrak{Z}_2' = \mathfrak{Z}_3 = \{\Omega\}$. This sequence is not distinguished. Note that the sequence

$$U_1 T_0 + V_1 T_1, \ U_2 T_0 T_1 + V_2 T_3, U_3 T_0 T_3 + V_3 T_2$$

is strictly admissible, as well as every sequence F_1, F_2, F_3 with the same pairs of reduced exponents M_1, M_2, M_3.

Strict admissibility of the sequence $F_1, \ldots, F_n$ of generic binomials can be characterized using the concepts introduced above.

8.10 Theorem *For a sequence $F_j = U_j T^{\mu(j)} + V_j T^{\nu(j)}$, $j = 1, \ldots, n$, of generic binomials in $Q[T] = \mathbb{Z}[U_1, V_1, \ldots, U_n, V_n][T_0, \ldots, T_n]$ and the corresponding sequence $M_j = (C_j, D_j)$, $j = 1, \ldots, n$, the following conditions are equivalent:*

(1) The sequence $F_1, \ldots, F_n$ is strictly admissible.

(2) There is a permutation π such that the sequence $M_{\pi 1}, \ldots, M_{\pi n}$ is distinguished.

(3) For an arbitrary noetherian ring $A \neq 0$ and arbitrary units u_j, v_j of A the sequence

$$f_1 := u_1 T^{\mu(1)} + v_1 T^{\nu(1)}, \ldots, f_n := u_n T^{\mu(n)} + v_n T^{\nu(n)}, T_0 \cdots T_n$$

in $A[T]$ and all its subsequences are regular.

(4) There are a noetherian ring $A \neq 0$ and elements a_j, b_j, c of A such that the sequence

$$a_1 T^{\mu(1)} + b_1 T^{\nu(1)}, \ldots, a_n T^{\mu(n)} + b_n T^{\nu(n)}, c T_0 \cdots T_n$$

in $A[T]$ is regular.

If these conditions are fulfilled, then the subgroup

$$\Gamma' := \mathbb{Z}(\mu(1) - \nu(1)) + \cdots + \mathbb{Z}(\mu(n) - \nu(n)) \subseteq \mathbb{Z}^{n+1}$$

is free of rank n and there is a grading of $Q[T]$ given by uniquely determined positive weights $\gamma = (\gamma_0, \ldots, \gamma_n)$ with $\gcd(\gamma_0, \ldots, \gamma_n) = 1$ such that the binomials $F_1, \ldots, F_n$ are homogeneous (of positive degrees) with respect to it and form a strongly regular sequence in all cases together with $T_0 \cdots T_n$.

Proof. (1) implies (2). We follow the line of reasoning in the proof of Lemma 6 in [11] and construct recursively the sequence $M_{\pi 1}, \ldots, M_{\pi n}$ in such a way that for the corresponding partitions $\mathfrak{Z}_0, \ldots, \mathfrak{Z}_n$ the sets $\mathfrak{Z}_{j-1}(C_{\pi j})$, $\mathfrak{Z}_{j-1}(D_{\pi j})$ are different and contain just one element of $\mathfrak{Z}_{j-1}$ each, $j = 1, \ldots, n$.

Assume that $M_{\pi 1}, \ldots, M_{\pi s}$ are already constructed, $0 \leq s < n$. After a renumbering we may assume in addition $\pi 1 = 1, \ldots, \pi s = s$. Then we must find an index $k \in \{s+1, \ldots, n\}$ such that $\mathfrak{Z}_s(C_k)$, $\mathfrak{Z}_s(D_k)$ are different and contain just one element of $\mathfrak{Z}_s$ each.

Consider an element $Z \in \mathfrak{Z}_s$ and let $m := |Z|$. Then the set

$$J := \{j \in \{1, \ldots, n\} : M_j \text{ rests on } \Omega \setminus Z\}$$

contains at most $|\Omega \setminus Z| - 1 = n - m$ elements because $F_1, \ldots, F_n$ is strictly admissible. Therefore, the complement

$$I := \{1, \ldots, n\} \setminus J = \{j \in \{1, \ldots, n\} : C_j \subseteq Z \text{ or } D_j \subseteq Z\}$$

contains at least m elements. There is an index $j(Z) \in I$ such that $M_{j(Z)}$ does not rest on Z. In case $C_{j(Z)} \subseteq Z$, $D_{j(Z)} \cap Z = \emptyset$; in case $D_{j(Z)} \subseteq Z$, $C_{j(Z)} \cap Z = \emptyset$.

Let $j \in I$ with $j \leq s$. Then $C_j \cup D_j$ is contained in some element of $\mathfrak{Z}_j$ and therefore in some element of $\mathfrak{Z}_s$, which must be Z. In particular, M_j rests on Z. This implies $j(Z) > s$.

Since $|\mathfrak{Z}_s| = n + 1 - s$, the mapping $\mathfrak{Z}_s \to \{s+1, \ldots, n\}$ by $Z \mapsto j(Z)$ is not injective. Therefore there are two different sets $X, Y \in \mathfrak{Z}_s$ with $j(X) = j(Y) =: k$. Obviously, k is an index of the kind we are looking for.

(2) implies (3). In a first step we prove that $A[T]/(f_1, \ldots, f_n, T_0 \cdots T_n)$ is a locally complete intersection of relative dimension 0 over A. For this we may assume by 6.2(4) that A is a field K. We have to prove now that $f_1, \ldots, f_n, T_0 \cdots T_n$ is a regular sequence in $K[T]$. This sequence is regular if and only if $f_1, \ldots, f_n, T_i$ is regular for every i. Furthermore we may assume that the sequence $M_1, \ldots, M_n$ itself is distinguished; let $\mathfrak{Z}_0, \ldots, \mathfrak{Z}_n$ be the corresponding sequence of partitions.

We show by induction on r that there exists a tuple $\gamma = (\gamma_0, \ldots, \gamma_n)$ of positive weights such that $f_1, \ldots, f_r, T_0 \cdots T_n$ are homogeneous and form a (strongly) regular sequence; we prove also that $\mu(1) - \nu(1), \ldots, \mu(r) - \nu(r)$ are linearly independent in $\mathbb{Z}^{n+1}$.

In the case $r = 0$, $T_0 \cdots T_n$ is a non-zero-divisor and homogeneous with respect to $\gamma = (1, \ldots, 1)$. Now assume that $f_1, \ldots, f_r, T_0 \cdots T_n$ are homogeneous with respect to $\tilde{\gamma}$ and form a regular sequence. There are (different) sets $X, Y \in \mathfrak{Z}_r$ such that $C_{r+1} \subseteq X$, $D_{r+1} \subseteq Y$. Let δ' and δ'' be the degrees of the monomials $T^{\mu(r+1)}$ and $T^{\nu(r+1)}$ with respect to the $\tilde{\gamma}$-grading and let $\gamma_i := \delta'' \tilde{\gamma}_i$ if $i \in X$, $\gamma_i := \delta' \tilde{\gamma}_i$ if $i \in Y$ and $\gamma_i := \tilde{\gamma}_i$ otherwise. Then f_{r+1} is homogeneous of degree $\delta' \delta''$ with respect to the γ-grading. The polynomials $f_1, \ldots, f_r$ are also homogeneous with respect to this grading, because for $j \leq r$ the set $C_j \cup D_j$ is contained in some set of $\mathfrak{Z}_r$ and therefore in X, Y or some other set of $\mathfrak{Z}_r$.

Let $J_X := \{j \in \{1, \ldots, r\} : C_j \cup D_j \subseteq X\}$. Then $|J_X| + 1 = |X|$, because X originates from $|X| - 1$ unions which are used to transform some $\mathfrak{Z}_{j-1}$ into $\mathfrak{Z}_j$. Analogously, $|J_Y| + 1 = |Y|$ for $J_Y := \{j \in \{1, \ldots, r\} : C_j \cup D_j \subseteq Y\}$. For $j \notin J_X \cup J_Y$, the monomials of f_j are contained in $K[T_i : i \notin X \cup Y]$. Therefore, using the fact that f_j, $j \notin J_X \cup J_Y$, together with any T_i, $i \notin X \cup Y$, form a regular sequence, it will be enough to show that $f_j, j \in J_X \cup J_Y, f_{r+1}, T_i$ is regular for every $i \in X \cup Y$, that is that the radical of the ideal $\mathfrak{A}$ generated by these elements in $K[T_i : i \in X \cup Y]$ is the homogeneous maximal ideal $\mathfrak{M}$. Assume $i \in Y$. Again by induction hypothesis the radical of the ideal generated by f_j, $j \in J_X$, and $T^{\mu(r+1)}$ in $K[T_i : i \in X]$ is the maximal homogeneous ideal $\mathfrak{M}_X$, and the radical of the ideal generated by f_j, $j \in J_Y$, and T_i in $K[T_i : i \in Y]$ is the maximal homogeneous ideal $\mathfrak{M}_Y$. Let $\mathfrak{P}$ be a prime ideal containing $\mathfrak{A}$. Then $\mathfrak{P}$ contains $\mathfrak{M}_Y$. Thus $T^{\nu(r+1)} \in \mathfrak{P}$. Hence $T^{\mu(r+1)} \in \mathfrak{P}$, $\mathfrak{P}$ contains $\mathfrak{M}_X$ and $\mathfrak{P} = \mathfrak{M}$. The tuple $\mu(r+1) - \nu(r+1)$ is independent of $\mu(j) - \nu(j)$, $j \in J_X \cup J_Y$ (and thus independent of $\mu(j) - \nu(j)$, $j = 1, \ldots, r$): A relation

$$0 = \sum_{j \in J_X \cup J_Y} a_j(\mu(j) - \nu(j)) + a(\mu(r+1) - \nu(r+1))$$

implies $0 = \sum_{j \in J_X} a_j(\mu(j) - \nu(j)) + a \cdot \mu(r+1)$ and

$$0 = \sum_{j \in J_X} a_j\langle \mu(j) - \nu(j), \tilde{\gamma}\rangle + a\langle \mu(r+1), \tilde{\gamma}\rangle = a\langle \mu(r+1), \tilde{\gamma}\rangle,$$

hence $a = 0$ because of $\langle \mu(r+1), \tilde{\gamma} \rangle > 0$. Then $a_j = 0$ for all $j \in J_X \cup J_Y$ by induction hypothesis, which completes the induction.

We have proved that $B = A[T]/(f_1, \ldots, f_n, T_0 \cdots T_n)$ is a locally complete intersection of relative dimension 0 and that there is a grading of $A[T]$ given by positive weights $\gamma = (\gamma_0, \ldots, \gamma_n)$ on the indeterminates such that $f_1, \ldots, f_n$ are homogeneous. Then by 6.12(2), or by 6.11 and 7.2(1), the sequence $f_1, \ldots, f_n, T_0 \cdots T_n$ is strongly regular. Any subsequence then is a regular sequence, too. This finishes the proof of $(2) \Rightarrow (3)$.

(4) is weaker than (3), and (4) implies (1) by 8.1(1'') and 8.9.

As we have seen in the proof of $(2) \Rightarrow (3)$ there is a grading of $Q[T]$ given by positive weights $\gamma_0, \ldots, \gamma_n$ for $T_0, \ldots, T_n$ such that $F_1, \ldots, F_n$ are homogeneous of positive degrees. That this grading is unique up to a common factor of $\gamma_0, \ldots, \gamma_n$ follows from the fact that the canonical syzygies $\mu(j) - \nu(j) \in \mathbb{Z}^{n+1}$, $j = 1, \ldots, n$, of $\gamma_0, \ldots, \gamma_n$ are linearly independent. $\square$

Theorem 8.10 holds verbatim for sequences of generic binomials $F_1, \ldots, F_r$ with $r \leq n$, one part excepted: In the case $r < n$ the gradings are not uniquely determined. This generalized version of 8.10 follows immediately from 8.10 and the following extension theorem, cf. [36], Theorem 3.1:

8.11 Theorem *Every strictly admissible sequence $F_1, \ldots, F_r$, $r \leq n$, of generic binomials in indeterminates $T_0, \ldots, T_n$ can be extended to a strictly admissible sequence $F_1, \ldots, F_r, \ldots, F_n$ of generic binomials in the same indeterminates $T_0, \ldots, T_n$.*

For a proof see [36], Section 3.

Of course, it is possible to extend any admissible (respectively strictly admissible) sequence of generic polynomials in $T_0, \ldots, T_n$ of length $\leq n + 1$ to an admissible (respectively strictly admissible) sequence of maximal length $n + 1$ simply by using additional generic linear forms $F_j = \sum_{i=0}^{n} U_{ij} T_i$. In the homogeneous case the additional generic forms can be chosen homogeneous, too, for instance by replacing the monomial T_i by T_i^{m/γ_i}, $m = \text{lcm}(\gamma_0, \ldots, \gamma_n)$ (where $\gamma_i = \deg T_i$).

The question arises whether the extension problem can be solved using generic binomials only. Apart from Theorem 8.11 this is not possible as the following examples show. The strictly admissible sequence

$$F_1 = U_1 T_0 + V_1 T_1 T_2, \quad F_2 = U_2 T_1 + V_2 T_2$$

cannot be extended to a strictly admissible sequence F_1, F_2, F_3 of generic binomials in T_0, T_1, T_2. The admissible sequence

$$F_1 = U_1 T_0 T_1 T_2 + V_1 T_0 T_1 T_3, \quad F_2 = U_2 T_0 T_2 T_3 + V_2 T_1 T_2 T_3$$

cannot be extended to an admissible sequence F_1, F_2, F_3 of generic binomials in T_0, T_1, T_2, T_3.

Supplements

1. Let $\gamma = (\gamma_0, \ldots, \gamma_n)$ and $\delta = (\delta_0, \ldots, \delta_r)$ consist of positive weights and degrees, respectively. Furthermore let non-empty subsets $N_j \subseteq \mathrm{N}_{\delta_j}(\gamma)$ be given and the corresponding generic homogeneous polynomials $F_j = \sum_{\nu \in N_j} U_{j\nu} T^\nu$, $j = 0, \ldots, r$. The following points discuss admissibility of $N_0, \ldots, N_r$ in terms of monoids generated by subsystems of the weights. For any subset $I \subseteq \{0, \ldots, n\}$ let $\mathrm{Mon}(\gamma_i : i \in I)$ denote the additive submonoid of $\mathbb{N}$ generated by γ_i, $i \in I$. Note: $\delta_j \in \mathrm{Mon}(\gamma_i : i \in I)$ if F_j does not belong to the ideal generated by T_i, $i \notin I$; the converse is true if the sets $N_0, \ldots, N_r$ are saturated.
a) If the sequence $N_0, \ldots, N_r$ is admissible then the following condition holds:
(R1) *For every non-empty proper subset I of $\{0, \ldots, n\}$ with $|I| \geq n - r$ the number of $j \in \{0, \ldots, r\}$ with $\delta_j \in \mathrm{Mon}(\gamma_i : i \in I)$ is $\geq |I| - (n - r)$.*
The converse holds if $N_0, \ldots, N_r$ are saturated.
b) Assume $r < n$. If the sequence $N_0, \ldots, N_r$ is strictly admissible then the following condition holds:
(R2L) *For every non-empty proper subset I of $\{0, \ldots, n\}$ with $|I| \geq n - r$ the number of $j \in \{0, \ldots, r\}$ with $\delta_j \in \mathrm{Mon}(\gamma_i : i \in I)$ is $> |I| - (n - r)$.*
The converse holds if $N_0, \ldots, N_r$ are saturated.
c) Assume $r = n$. If the sequence $N_0, \ldots, N_n$ is strictly admissible then the following condition holds:
(R2M) *For every non-empty proper subset I of $\{0, \ldots, n\}$ the number of $j \in \{0, \ldots, n\}$ with $\delta_j \in \mathrm{Mon}(\gamma_i : i \in I)$ is $> |I|$.*
The converse holds if $N_0, \ldots, N_n$ are saturated.
(Remark. In the classical case $\gamma_0 = \cdots = \gamma_n = 1$ the three conditions (R1), (R2L) and (R2M) are trivially fulfilled for arbitrary positive degrees.)
d) Assume $r = n$. If for every j the degree δ_j is a common multiple of the weights $\gamma_0, \ldots, \hat{\gamma}_j, \ldots, \gamma_n$ (γ_j omitted) then (R2M) is fulfilled. The same is true if the δ_j have both of the following properties (used by J.P. Jouanolou in Section 6.3.10 of [22]): (H_1) For every i the weight γ_i divides all the δ_j with the possible exception of one of them. (H_2) For every j and every i, $\delta_j \in \mathrm{Mon}(\gamma_s : s \neq i)$.

2.† Let $F_0, \ldots, F_r$, $r \leq n$, be an admissible sequence of homogeneous generic polynomials in $T_0, \ldots, T_n$. In the polynomial ring $Q_{(k)}[T] \otimes_{Q_{(k)}} Q_{(k)}[T]$ we denote the indeterminates $T_i \otimes 1$ by T_i and $1 \otimes T_i$ by T_i'. Similarly, we set $F_j := F_j \otimes 1$ and $F_j' := 1 \otimes F_j$. Let $f_j := F_j - F_j'$.
a) $f_0, \ldots, f_r$ form a regular sequence in $Q_{(k)}[T, T']$ for every noetherian ring $k \neq 0$. (Sect. 7, Suppl. 6a).)
b) Assume $r < n$ and let the sequence $F_0, \ldots, F_r$ be strictly admissible. Then $f_0, \ldots, f_r$ generate a prime ideal of codimension $r + 1$ in $Q_{(k)}[T, T']$ for every noetherian integral domain k. (Sect. 7, Suppl. 6b).)

9 The Generic Structure of the Principal Component

Let $F_0, \ldots, F_r$ be generic polynomials $\neq 0$ in $Q_{(k)}[T_0, \ldots, T_n]$, k a noetherian integral domain. In this section we study the field extension corresponding to the inclusion

$$Q_{(k)}/\mathfrak{H}^\star \cap Q_{(k)} \longrightarrow H_{(k)} = Q_{(k)}[T, T^{-1}]/(F_0, \ldots, F_r)$$

of integral domains, where $\mathfrak{H}^\star$ is the principal component of $(F_0, \ldots, F_r)$ introduced in the last section.

We start with the following proposition:

9.1 Proposition *Let $F_0, \ldots, F_r$ be a sequence of generic polynomials $\neq 0$ in $Q_{(k)}[T]$, k a noetherian integral domain. Assume that the canonical homomorphism $Q_{(k)} \to Q_{(k)}[T, T^{-1}]/(F_1, \ldots, F_r)$ is injective. Then the height of the prime ideal*

$$\mathfrak{H}^\star \cap Q_{(k)} \subseteq Q_{(k)}$$

is ≤ 1. In particular, this is true if $r \leq n$ and the sequence $F_1, \ldots, F_r$ is strictly admissible, i. e. if $F_1, \ldots, F_r$ generate a prime ideal in $Q_{(k)}[T]$.

Proof. Going over to the quotient field of k, we may assume right away that k itself is a field.

$\mathfrak{H}^\star \cap Q_{(k)}$ is the kernel of the canonical homomorphism from $Q_{(k)}$ to $H_{(k)} := Q_{(k)}[T, T^{-1}]/(F_0, \ldots, F_r)$. Thus it will be enough to show that its image has transcendence degree $\geq \dim Q_{(k)} - 1$ over k. Let $Q'_{(k)} \subseteq Q_{(k)}$ denote the k-subalgebra generated by all the $U_{j\nu}$ with the exception of one of the $U_{0\nu}$, arbitrarily chosen among the coefficients of F_0. We will prove that $Q'_{(k)} \to H_{(k)}$ is injective. Indeed, in the canonical sequence

$$Q'_{(k)} \xrightarrow{\alpha_0} Q'_{(k)}[T]/(F_1, \ldots, F_r) \xrightarrow{\alpha_1} Q'_{(k)}[T, T^{-1}]/(F_1, \ldots, F_r) \xrightarrow{\alpha_2} H_{(k)}$$

the mapping α_2 is an isomorphism and the composition $\alpha_1 \circ \alpha_0$ is injective (which is equivalent to the injectivity of $Q_{(k)} \to Q_{(k)}[T, T^{-1}]/(F_1, \ldots, F_r)$). The additional statement follows this way: If $(F_1, \ldots, F_r)$ is a prime ideal (not containing any of the indeterminates $T_0, \ldots, T_n$), then even α_1 is injective. (Let us note that the composition $\alpha_1 \circ \alpha_0$ may be injective without α_1 being injective. For an example see Supplement 3.) $\qquad\square$

If in the situation of 9.1 k is a factorial domain, then $\mathfrak{H}^\star \cap Q_{(k)}$ is a principal ideal. It may be the zero ideal, though, as is shown by the following (strictly admissible) example: $F_0 = U_1 T_0 + U_2 T_0^2 + U_3 T_1$, $F_1 = V_1 T_0 + V_2 T_1$. Here $\mathfrak{T}^\star = \mathfrak{H}^\star$ by 8.8 and $\mathfrak{T}^\star \cap Q_{(k)} = \mathfrak{H}^\star \cap Q_{(k)} = 0$. This example shows that for the theory of resultants to work one has to specialize the situation e. g. specialize to the case of homogeneous polynomials.

Our further results on the generic structure of the principal component are based on the next lemma:

9.2 Main Lemma *Let K be a field of characteristic 0, $M_1, \ldots, M_r$ finite subsets of $\mathbb{Z}^m$ and $\Gamma_1, \ldots, \Gamma_r$ the subgroups of $\mathbb{Z}^m$ generated by them. Furthermore, let*

$$G_j := \sum_{\alpha \in M_j} U_{j\alpha} X^\alpha \in L[X, X^{-1}]$$

be Laurent polynomials in $X_1, \ldots, X_m$ with indeterminate coefficients over the field $L := K(U_{j\alpha})_{j,\alpha}$. Then the following conditions are equivalent:

(1) $G_1, \ldots, G_r$ *are algebraically independent over L.*

(2) *For every subset $J \subseteq \{1, \ldots, r\}$ the group $\Gamma_J := \sum_{j \in J} \Gamma_j \subseteq \mathbb{Z}^m$ has rank $\geq |J|$.*

If for every $i \in \{1, \ldots, r\}$ the $r - 1$ polynomials G_j, $j \neq i$, are algebraically independent over L, but $G_1, \ldots, G_r$ in total are algebraically dependent over L, then

$$L(G_1, \ldots, G_r) = L(X^\alpha : \alpha \in \Gamma_1 + \cdots + \Gamma_r),$$

i. e. the field extension $L \subseteq L(G_1, \ldots, G_r)$ is purely transcendental. If $\alpha_1, \ldots, \alpha_{r-1}$ form a $\mathbb{Z}$-base of $\Gamma_1 + \cdots + \Gamma_r$ then $X^{\alpha_1}, \ldots, X^{\alpha_{r-1}}$ is a transcendence base. Furthermore, for every i the degree of the field extension $L(G_j : j \neq i) \subseteq L(G_1, \ldots, G_r)$ is divisible by the group index $[(\Gamma_1 + \cdots + \Gamma_r) : \sum_{j \neq i} \Gamma_j]$.

Proof. In general, the inclusion $L(G_j : j \in J) \subseteq L(X^\alpha : \alpha \in \Gamma_J)$ holds for every subset $J \subseteq \{1, \ldots, r\}$. Thus (1) implies (2) .

The proof of (2) $\Rightarrow$ (1) uses induction on $|J|$. Therefore it suffices to show the last assertion of the lemma. For this we have to show that for every $j = 1, \ldots, r$ the monomials X^α, $\alpha \in \Gamma_j$, belong to $L(G_1, \ldots, G_r)$. We demonstrate this for the case $j = r$. More precisely:

9.3 Lemma *In the situation of 9.2 let $G_1, \ldots, G_{r-1}$ be algebraically independent and $G_1, \ldots, G_{r-1}, G_r$ be algebraically dependent over L. Then*

$$L(G_1, \ldots, G_{r-1}, G_r) = L(G_1, \ldots, G_{r-1}; X^\alpha : \alpha \in M_r).$$

The degree of the (finite) field extension

$$L(G_1, \ldots, G_{r-1}) \subseteq L(G_1, \ldots, G_{r-1}, G_r)$$

is less than or equal to the (finite) index $[(\Gamma_1 + \cdots + \Gamma_r) : (\Gamma_1 + \cdots + \Gamma_{r-1})]$.

Proof of 9.3. There is a minimal equation $H(G_r) = 0$ with a prime polynomial

$$H = h_0 + h_1 Y + \cdots + h_s Y^s \in K[U_{j\alpha}][G_1, \ldots, G_{r-1}][Y].$$

First we show that all monomials X^α, $\alpha \in M_r$, are algebraic over the ring $K[U_{j\alpha}][G_1, \ldots, G_{r-1}]$ which is equivalent with $G_\alpha := G_r - U_{r\alpha}X^\alpha$ being algebraic for every $\alpha \in M_r$. But, one gets G_α from G_r by specializing $U_{r\alpha} \mapsto 0$ and the polynomial H remains non-trivial under this specialization because its coefficients do not have $U_{r\alpha}$ as a common divisor.

The X^α, $\alpha \in M_r$, are also algebraic over $K[U_{j\alpha} : j < r][G_1, \ldots, G_{r-1}]$. By Theorem 2.7 applied to the simple (finite and separable) field extension

$$K(U_{j\alpha, j<r})(G_1, \ldots, G_{r-1}) \subseteq K(U_{j\alpha, j<r})(G_1, \ldots, G_{r-1})(X^\alpha : \alpha \in M_r)$$

we obtain that the field $L(G_1, \ldots, G_{r-1}, G_r)$ actually contains all the X^α, $\alpha \in M_r$. □

The additional statement of Lemma 9.2 about degrees of field extensions follows from the inclusions

$$L(G_j : j \neq i) \subseteq L(X^\alpha : \alpha \in \textstyle\sum_{j \neq i} \Gamma_j) \subseteq L(X^\alpha : \alpha \in \Gamma_1 + \cdots + \Gamma_r)$$

and the fact that in each case the degree of the second extension coincides with the group index specified. □

9.4 Corollary *In the situation of Lemma 9.2 let $r = m$. Then the field extension $L(G_1, \ldots, G_m) \subseteq L(X_1, \ldots, X_m)$ is algebraic if and only if for every $J \subseteq \{1, \ldots, m\}$ rank$\Gamma_J \geq |J|$. In this case its degree is divisible by the group index $[\mathbb{Z}^m : (\Gamma_1 + \cdots + \Gamma_m)]$.*

For the *proof* note that $L(G_1, \ldots, G_m) \subseteq L(X_1, \ldots, X_m)$ is algebraic if and only if the elements $G_1, \ldots, G_m$ are algebraically independent over L. The inclusions

$$L(G_1, \ldots, G_m) \subseteq L(X^\alpha : \alpha \in \Gamma_1 + \cdots + \Gamma_m) \subseteq L(X_1, \ldots, X_m)$$

yield the additional statement about the group index. □

We come back to the discussion of the principal component. For this, the following notation will be very useful.

Definition For every $N \subseteq \mathbb{N}^{n+1}$ let $\Gamma(N)$ denote the subgroup

$$\Gamma(N) := \sum_{\nu, \mu \in N} \mathbb{Z}(\nu - \mu) \subseteq \mathbb{Z}^{n+1}.$$

Now let $F_0, \ldots, F_r$ again be generic polynomials $\neq 0$. For every $j \in \{0, \ldots, r\}$ let Γ_j denote the group

$$\Gamma_j := \Gamma(N_j).$$

In $N_j(\neq \emptyset)$ we fix an exponent

$$\nu(j) \in N_j.$$

Then Γ_j is already generated by the differences $\nu - \nu(j)$, $\nu \in N_j'$, where

$$N_j' := N_j \setminus \{\nu(j)\}.$$

Thus Γ_j is generated by the exponents of the Laurent polynomial

$$G_j := -\sum_{\nu \in N_j'} U_{j\nu} T^{\nu - \nu(j)}.$$

We have $F_j = T^{\nu(j)}(U_{j\nu(j)} - G_j)$ in $Q[T, T^{-1}]$. Let $Q_{(k)}^\flat$ denote the polynomial ring

$$Q_{(k)}^\flat := k[U_{j\nu} : j = 0, \ldots, r, \ \nu \in N_j'].$$

Then $Q_{(k)} = Q_{(k)}^\flat[U_{0\nu(0)}, \ldots, U_{r\nu(r)}]$ and there are canonical isomorphisms

$$
\begin{aligned}
H_{(k)} &= Q_{(k)}[T, T^{-1}]/(F_0, \ldots, F_r) \\
&= Q_{(k)}^\flat[T, T^{-1}][U_{0\nu(0)}, \ldots, U_{r\nu(r)}]/(U_{0\nu(0)} - G_0, \ldots, U_{r\nu(r)} - G_r) \\
&= Q_{(k)}^\flat[T, T^{-1}]
\end{aligned}
$$

which carry $U_{j\nu(j)}$ to G_j. In particular, the image of $Q_{(k)}$ in $H_{(k)}$ can be identified with $Q_{(k)}^\flat[G_0, \ldots, G_r]$ in $Q_{(k)}^\flat[T, T^{-1}]$.

The kernel $\mathfrak{H}^* \cap Q_{(k)}$ of $Q_{(k)} \to H_{(k)}$ vanishes if and only if $G_0, \ldots, G_r$ are algebraically independent over $Q_{(k)}^\flat$.

If k is an integral domain, the function field of the principal component is $L(T) = L(T_0, \ldots, T_n)$, where L is the quotient field of $Q_{(k)}^\flat$, and the function field of the integral domain $Q_{(k)}/\mathfrak{H}^* \cap Q_{(k)}$ is $L(G_0, \ldots, G_r) \subseteq L(T_0, \ldots, T_n)$. Applying 9.2 and 9.4 we get:

9.5 Theorem *Let k be a noetherian integral domain of characteristic 0. In the situation just described, the following conditions are equivalent:*
(1)   The mapping $Q_{(k)} \to H_{(k)}$ is injective.
(2)   For every $J \subseteq \{0, \ldots, r\}$ the group $\Gamma_J := \sum_{j \in J} \Gamma_j$ has rank $\geq |J|$.
Let these conditions be fulfilled. Then $r \leq n$. If $r = n$ then the (finite) degree of the quotient field of $H_{(k)}$ over the quotient field of $Q_{(k)}$ is divisible by the group index $[\mathbb{Z}^{n+1} : (\Gamma_0 + \cdots + \Gamma_n)]$.

Since for a strictly admissible sequence $F_0, \ldots, F_r$ in $Q_{(k)}[T]$ with $r < n$ the ideal $\mathfrak{H}^* = (F_0, \ldots, F_r)$ is prime and the canonical homomorphism $Q_{(k)} \to H_{(k)}$ injective, we get as a corollary of 9.5:

9.6 Lemma *Assume $r < n$. If the sequence of non-empty finite subsets $N_0, \ldots, N_r$ of $\mathbb{N}^{n+1} \setminus \{0\}$ is strictly admissible, then for every subset $J \subseteq \{0, \ldots, r\}$ the group $\Gamma_J = \sum_{j \in J} \Gamma_j$ has rank $\geq |J|$.*

Lemma 9.6 yields only a necessary condition on strict admissibility as is shown simply by $r = 0$, $n = 1$ and $N_0 = \{(2,0),(1,1)\}$.

Furthermore we remark that $G_0, \ldots, G_r$ may be algebraically independent even in cases where the admissible polynomials $F_0, \ldots, F_r$ are not strictly admissible, as is shown simply by $r = 0$, $F_0 = U_1 T_0^2 + U_2 T_0 T_1$. See also Supplement 3.

As other applications of the Main Lemma 9.2 we are going to describe now two cases in which the map $Q_{(k)} \to H_{(k)}$ is not injective but has nonetheless important properties.

9.7 Theorem *Let k be a noetherian domain of characteristic 0. Let $r = n + 1$. Assume that for every proper subset J of $\{0, \dots, n + 1\}$ the group $\Gamma_J \subseteq \mathbb{Z}^{n+1}$ has rank $\geq |J|$. Then the following holds:*

The map $Q_{(k)} \to H_{(k)}$ is not injective, its kernel $\mathfrak{H}^\star \cap Q_{(k)}$ has codimension 1 and its image has the same quotient field as the group ring $Q_{(k)}^\triangleright[T^\alpha : \alpha \in \Gamma_0 + \cdots + \Gamma_{n+1}]$.

The quotient field of $Q_{(k)}/\mathfrak{H}^\star \cap Q_{(k)}$ is purely transcendental over the quotient field of k, a transcendence base, for example, consisting of the indeterminates $U_{j\nu}$, $j = 0, \dots, n+1$, $\nu \in N_j'$, together with the monomials $T^{\alpha(1)}, \dots, T^{\alpha(n+1)}$, where the exponents $\alpha(1), \dots, \alpha(n+1)$ form a $\mathbb{Z}$-base of $\Gamma_0 + \cdots + \Gamma_{n+1}$. In particular, the degree of the quotient field of $H_{(k)}$ over the quotient field of $Q_{(k)}/\mathfrak{H}^\star \cap Q_{(k)}$ is the group index $[\mathbb{Z}^{n+1} : (\Gamma_0 + \cdots + \Gamma_{n+1})]$.

9.8 Theorem *Let k be a noetherian domain of characteristic 0. Let $r = n$. Assume that for every proper subset J of $\{0, \dots, n\}$ the group Γ_J has rank $\geq |J|$, but $\Gamma_0 + \cdots + \Gamma_n$ has rank n (smaller than $|\{0, \dots, n\}|$). Then the following holds:*

The map $Q_{(k)} \to H_{(k)}$ is not injective, its kernel $\mathfrak{H}^\star \cap Q_{(k)}$ has codimension 1 and its image has the same quotient field as the group ring $Q_{(k)}^\triangleright[T^\alpha : \alpha \in \Gamma_0 + \cdots + \Gamma_n]$.

The quotient field of $Q_{(k)}/\mathfrak{H}^\star \cap Q_{(k)}$ is purely transcendental over the quotient field of k. A transcendence base of this field extension is formed, for example, by the indeterminates $U_{j\nu}$, $j = 0, \dots, n$, $\nu \in N_j'$, together with the monomials $T^{\alpha(1)}, \dots, T^{\alpha(n)}$ where the exponents $\alpha(1), \dots, \alpha(n)$ form a $\mathbb{Z}$-base of $\Gamma_0 + \cdots + \Gamma_n$.

9.9 Corollary *In the situation of Theorem 9.7 or Theorem 9.8, if k and hence $Q_{(k)}$ is a factorial domain, the ideal $\mathfrak{H}^\star \cap Q_{(k)}$ is a principal prime ideal $\neq 0$.*

The assumptions of Theorem 9.8 are satisfied in the following situation:

9.10 Proposition *Let $F_0, \dots, F_n$ be a strictly admissible sequence of generic polynomials and let k be a noetherian integral domain of characteristic 0. If $\mathfrak{H}^\star \cap Q_{(k)} \neq 0$, then $\operatorname{rank} \Gamma_J \geq |J|$ for every proper subset J of $\{0, \dots, n\}$ and $\operatorname{rank}(\Gamma_0 + \cdots + \Gamma_n) = n$.*

Proof. The part about $\operatorname{rank}\Gamma_J \geq |J|$ is covered by Lemma 9.6. Because of $\mathfrak{H}^\star \cap Q_{(k)} \neq 0$, the elements $G_0, \ldots, G_n$ (corresponding to $F_0, \ldots, F_n$) are algebraically dependent over $Q_{(k)}^\flat$. By 9.2, $\operatorname{rank}(\Gamma_0 + \cdots + \Gamma_n) = n$. $\square$

Supplements

1.† Let $F_0, \ldots, F_r$ be an admissible sequence of generic polynomials in $T_0, \ldots, T_n$ and let k be an arbitrary noetherian integral domain of characteristic 0. Then the codimension of the prime ideal $\mathfrak{H}^\star \cap Q_{(k)}$ is the corank of the Jacobian matrix $(\partial(F_0, \ldots, F_n)/\partial(T_0, \ldots, T_n))$ considered over the integral domain $H_{(k)} = Q_{(k)}[T, T^{-1}]/(F_0, \ldots, F_n)$, i.e. the rank of the differential module $\Omega_{Q_{(k)}}(H_{(k)})$.

2. Under the assumptions of Proposition 9.1 assume in addition that k is of characteristic 0 and that $\mathfrak{H}^\star \cap Q_{(k)}$ is of codimension 1. Then the quotient field of the image of $Q_{(k)}$ in $H_{(k)}$ contains all the (residue classes of) T^α, $\alpha \in M_0 = \{\nu - \mu : \nu, \mu \in N_0\}$. (Lemma 9.3.)

3. The sequence $F_0 := U_1 T_0 T_2 + U_2 T_0^2 + U_3 T_1 T_2$, $F_1 := V_1 T_0 + V_2 T_1$ is admissible but not strictly admissible. For any integral domain k, the homomorphism $Q_{(k)} \to Q_{(k)}[T, T^{-1}]/(F_0, F_1)$ is injective and the corresponding Laurent polynomials $G_0 = U_3 - T_1^{-1} T_2^{-1} F_0$, $G_1 = V_2 - T_1^{-1} F_1$ are algebraically independent over the quotient field of $k[U_1, U_2, V_1]$.

III Elimination

10 Basics of Elimination

We consider a noetherian (positively) graded ring

$$C = \bigoplus_{m \geq 0} C_m.$$

Then C is a finitely generated algebra over the noetherian ground ring $A = C_0$. There is the decreasing sequence

$$C_{(k)} := \bigoplus_{m \geq k} C_m, \quad k \in \mathbb{N},$$

of homogeneous ideals;

$$C_+ := C_{(1)}$$

is the irrelevant ideal of C. The projective scheme $\operatorname{Proj} C$ is noetherian and of finite type over the affine scheme $\operatorname{Spec} A$. Let

$$X = \operatorname{Proj} C \xrightarrow{h} \operatorname{Spec} A$$

denote the structure morphism. X is covered by the affine open sets

$$D_+(f) = \operatorname{Spec} (C_f)_0$$

where $f \in C_+$ is homogeneous and $(C_f)_0$ is the homogeneous part of degree 0 of the graded ring of fractions C_f. The points of X are the homogeneous prime ideals $\mathfrak{P}$ in C not containing C_+. The image $h(\mathfrak{P})$ is the prime ideal $\mathfrak{P} \cap A = \mathfrak{P}_0$ in A, that is the homogeneous part of degree 0 of $\mathfrak{P}$. *The set X is empty if and only if C_+ is nilpotent, i.e. $C_m = 0$ for m big enough, or equivalently: C is finite over A.*

The fundamental problem of elimination is to determine the image of h in $\operatorname{Spec} A$. To do this we define the elimination ideal $\mathfrak{T}_0(C) \subseteq A$ by

$$\mathfrak{T}_0(C) = \bigcup_{k \geq 0} \operatorname{Ann}_A C_{(k)} = \bigcup_{k \geq 0} \bigcap_{m \geq k} \operatorname{Ann}_A C_m.$$

The ideals $\operatorname{Ann}_A C_{(k)}$ in A form an increasing sequence. If $C = A[t_0, \ldots, t_n]$ with homogeneous elements $t_0, \ldots, t_n$ of positive degrees $\gamma_0, \ldots, \gamma_n$, the maximum of which we shall denote by μ, then obviously

$$\operatorname{Ann}_A C_{(k)} = \bigcap_{m=k}^{k+\mu-1} \operatorname{Ann}_A C_m = \operatorname{Ann}_A \bigoplus_{m=k}^{k+\mu-1} C_m$$

is the annihilator of a finite A-module. Thus, for a prime ideal $\mathfrak{p} \in \operatorname{Spec} A$ the following three conditions are equivalent:

(1) $\mathfrak{p} \notin V(\operatorname{Ann}_A C_{(k)})$.   (2) $\bigoplus_{m=k}^{k+\mu-1} (C_m)_\mathfrak{p} = 0$.   (3) $(C_\mathfrak{p})_{(k)} = (C_{(k)})_\mathfrak{p} = 0$.

10.1 Main Theorem of Elimination

$$\operatorname{Im}(h) \;=\; V(\mathfrak{T}_0(C)).$$

In particular, the image of h is a closed set in $\operatorname{Spec} A$.

Proof. For $\mathfrak{p} \in \operatorname{Spec} A$, by the last remark, the following are equivalent:
(1) $\mathfrak{p} \notin \operatorname{Im}(h)$.   (2) $\emptyset = h^{-1}(\mathfrak{p}) = \operatorname{Proj}(C_\mathfrak{p}/\mathfrak{p}C_\mathfrak{p})$.   (3) $(C_\mathfrak{p}/\mathfrak{p}C_\mathfrak{p})_{(k)} = 0$ for some k.   (4) $(C_\mathfrak{p})_{(k)} = 0$ for some k.   (5) $\mathfrak{p} \notin V(\operatorname{Ann}_A C_{(k)})$ for some k.
(6) $\mathfrak{p} \notin V(\mathfrak{T}_0(C))$.   □

If $\mathfrak{A} = \bigoplus_{m \geq 0} \mathfrak{A}_m$ is a homogeneous ideal in C then the elimination ideal $\mathfrak{T}_0(C/\mathfrak{A}) \subseteq C_0/\mathfrak{A}_0 = A/\mathfrak{A}_0$ is often called the e l i m i n a t i o n   i d e a l   o f $\mathfrak{A}$. The image of the closed subvariety $\operatorname{Proj}(C/\mathfrak{A}) = V_+(\mathfrak{A}) \subseteq \operatorname{Proj} C$ with respect to the structure morphism $h : \operatorname{Proj} C \to \operatorname{Spec} A$ is the closed set $V(\mathfrak{T}_0(C/\mathfrak{A})) \subseteq \operatorname{Spec}(A/\mathfrak{A}_0) = V(\mathfrak{A}_0) \subseteq \operatorname{Spec} A$. In particular, *the mapping $h : \operatorname{Proj} C \to \operatorname{Spec} A$ is closed.*

10.2 Proposition *Let $A \to A'$ be any change of noetherian ground rings and $C' := A' \otimes_A C = \bigoplus_{m \geq 0} A' \otimes_A C_m = \bigoplus_{m \geq 0} C'_m$. Then*

$$\mathfrak{T}_0(C)A' \;\subseteq\; \mathfrak{T}_0(C') \;\subseteq\; \sqrt{\mathfrak{T}_0(C)A'}.$$

Proof. Both inclusions stem from the fact that for any finite A-module V the inclusions $(\operatorname{Ann}_A V)A' \subseteq \operatorname{Ann}_{A'}(A' \otimes V) \subseteq \sqrt{(\operatorname{Ann}_A V)A'}$ hold.   □

In the situation of Proposition 10.2 we have

$$\operatorname{Proj} C' \;=\; A' \otimes_A \operatorname{Proj} C \;=\; \operatorname{Spec} A' \times_{\operatorname{Spec} A} \operatorname{Proj} C.$$

That both ideals $\mathfrak{T}_0(C)A'$ and $\mathfrak{T}_0(C')$ have the same radical then follows also by 10.1 using the Cartesian diagram

$$
\begin{array}{ccc}
\operatorname{Proj} C' & \longrightarrow & \operatorname{Proj} C \\
\downarrow & & \downarrow \\
\operatorname{Spec} A' & \longrightarrow & \operatorname{Spec} A \,.
\end{array}
$$

The elimination ideal is but the homogeneous part of degree 0 of the ideal of i n e r t i a   f o r m s (called T r ä g h e i t s f o r m e n by Hurwitz) of C:

$$\mathfrak{T}(C) \;=\; \bigcup_{k \geq 0} \operatorname{Ann}_C C_{(k)}$$

which is the ideal of elements in C with support in $V(C_+) \subseteq \operatorname{Spec} C$, see [19], for instance. The canonical embedding $\operatorname{Proj}(C/\mathfrak{T}(C)) \to \operatorname{Proj} C$ is a homeomorphism. If

$$0 = \mathfrak{Q}_1 \cap \ldots \cap \mathfrak{Q}_r$$

is a Lasker-Noether decomposition with homogeneous primary ideals $\mathfrak{Q}_i$ such that $C_+ \not\subseteq \sqrt{\mathfrak{Q}_i}$ exactly for $i \leq r_0$, then

$$\mathfrak{T}(C) = \mathfrak{Q}_1 \cap \ldots \cap \mathfrak{Q}_{r_0}.$$

In addition, $\mathfrak{T}(C) = \operatorname{Ann}_C \mathfrak{Q} = \operatorname{Hom}_C(C/\mathfrak{Q}, C)$ with

$$\mathfrak{Q} = \mathfrak{Q}(C) := \mathfrak{Q}_{r_0+1} \cap \ldots \cap \mathfrak{Q}_r.$$

(*Proof.* Consider an element of $\mathfrak{Q}$ relatively prime to $\mathfrak{T}(C)$.)

10.3 Proposition *Let $A \to A'$ be a flat homomorphism of noetherian rings and $C' := A' \otimes_A C$. Then*

$$\mathfrak{T}(C') = \mathfrak{T}(C)C' = \mathfrak{T}(C)A'$$

and in particular $\mathfrak{T}_0(A' \otimes_A C) = \mathfrak{T}_0(C)A'$.

Proof. The homomorphism $C \to C'$ is flat, too. Therefore $\operatorname{Ann}_{C'} C'_{(k)} = C' \operatorname{Ann}_C C_{(k)}$ for every $k \in \mathbb{N}$. □

Any noetherian graded A-algebra $C = \bigoplus_{m \geq 0} C_m$ (with $C_0 = A$) is obtained by a specialization $Q \to A$ of an algebra of type

$$S = \bigoplus_{m \geq 0} S_m = Q[T_0, \ldots, T_n]/(F_0, \ldots, F_r)$$

where $T_0, \ldots, T_n$ are indeterminates of positive degrees, say $\gamma_0, \ldots, \gamma_n$, and where $F_0, \ldots, F_r$ are generic homogeneous polynomials of positive degrees $\delta_0, \ldots, \delta_r$:

$$F_j = \sum_{\nu_0 \gamma_0 + \cdots + \nu_n \gamma_n = \delta_j} U_{j\nu} T^\nu,$$

$j = 0, \ldots, r$. The ground ring $Q = S_0$ is the polynomial ring

$$Q := \mathbb{Z}[U_{j\nu}]_{j,\nu}.$$

Sometimes the ring $\mathbb{Z}$ in Q will be replaced by another noetherian ring k, e. g. a field. In that case we will use the notation $Q_{(k)}$ instead of Q.

The elimination ideal $\mathfrak{T}_0(S)$ in Q is called the **generic elimination ideal**. For any specialization $C = A \otimes_Q S$ the A-ideal $\mathfrak{T}_0(S) \cdot A$ is called the **elimination ideal corresponding to the representation** of C. In general it differs from the elimination ideal $\mathfrak{T}_0(C) \subseteq A$ but by 10.2 it has the same set of zeros in $\operatorname{Spec} A$.

For $r < n$ the generic elimination ideal is the zero ideal, obviously (since $\mathrm{Proj}\,(K \otimes_Q S) \neq \emptyset$, i.e. $\mathfrak{I}_0(S) \cdot K = 0$, K the quotient field of Q). In this case the elimination ideal $\mathfrak{I}_0(C)$ is nilpotent. The first interesting case $r = n$ is the so-called **main case of elimination**.

For a torsionfree $\mathbb{Z}$-algebra k the generic elimination ideal defined in $Q_{(k)} = k \otimes_{\mathbb{Z}} Q$ coincides with the elimination ideal $\mathfrak{I}_0(S)Q_{(k)}$. The case of arbitrary $\mathbb{Z}$-algebras seems to be unsettled.

Supplements

1. Let $C = \bigoplus_{m \geq 0} C_m$ be a noetherian graded A-algebra with $A := C_0$ and let $f_1, \ldots, f_s \in C_+$ homogeneous elements such that C_+ is contained in the radical of the ideal generated by $f_1, \ldots, f_s$. Then the ideal $\mathfrak{I}(C)$ of the inertia forms is the kernel of the canonical homomorphism
$$C \;\rightarrow\; \prod_{i=1}^s C_{f_i}\,.$$
In particular, the elimination ideal $\mathfrak{I}_0(C)$ is the kernel of the homomorphism
$$A \;\rightarrow\; \prod_{i=1}^s (C_{f_i})_0$$
in degree 0, which is the same as the kernel of the structure homomorphism $A \to \Gamma(\mathrm{Proj}\,C)$. (More generally, the A-module $\mathfrak{I}_m(C)$ of inertia forms of degree m is the kernel of the canonical homomorphism $C_m \to \Gamma(\mathcal{O}(m))$, where $\mathcal{O}(m)$ is the $\mathcal{O}$-module belonging to the shifted graded C-module $C(m)$, $\mathcal{O} := \mathcal{O}_{\mathrm{Proj}\,C}$.)

2. For $A := k[u, v]$ with $uv = 0, u^2 = v^2 \neq 0$, the elimination ideal $\mathfrak{I}_0(C)$ of the A-algebra $C := A[T_0, T_1]/(uT_0^d + vT_1^d)$, $d > 0$, $\deg T_0 = \deg T_1 = 1$, is $\neq 0$.

3. For an arbitrary noetherian ring k the generic elimination ideal in $Q_{(k)}$ is contained in the ideal generated by the indeterminates $U_{j\nu}$. (Specialize $U_{j\nu} \mapsto 0$.)

11 The Main Case for Generic Regular Sequences

In this section we start with the investigation of the main case of elimination in the generic situation.

Let $\gamma = (\gamma_0, \ldots, \gamma_n) \in (\mathbb{N}^*)^{n+1}$ be a given system of weights, furthermore $\delta = (\delta_0, \ldots, \delta_n) \in (\mathbb{N}^*)^{n+1}$ a given system of degrees and $N_j \subseteq N_{\delta_j}(\gamma)$ defining sets of exponents for the homogeneous generic polynomials $F_j = \sum_{\nu \in N_j} U_{j\nu} T^\nu$, $j = 0, \ldots, n$. Admissibility of $F_0, \ldots, F_n$ has been discussed in Section 8 earlier. In the context of elimination theory admissibility can be described using the graded algebra

$$S = \bigoplus_{m \geq 0} S_m = Q[T_0, \ldots, T_n]/(F_0, \ldots, F_n)$$

over $Q = \mathbb{Z}[U_{j\nu}]_{j,\nu}$ in the following way:

11.1 Lemma *The homogeneous generic polynomials $F_0, \ldots, F_n$ establish an admissible sequence if and only if the elimination ideal $\mathfrak{T}_0(S)$ of the Q-algebra $S = Q[T]/(F_0, \ldots, F_n)$ is not the zero ideal.*

Proof. $\mathfrak{T}_0(S)$ is different from zero if and only if the morphism $\mathrm{Proj}\, S \to \mathrm{Spec}\, Q$ is not surjective. This in turn is equivalent to the fact that the scheme $\mathrm{Proj}\,(L[T]/(F_0, \ldots, F_n))$ is empty, $L := \mathbb{Q}(U_{j\nu})_{j,\nu}$. But this characterizes the admissibility of γ, δ by Lemma 8.1, conditions (2),(3). $\quad\square$

As a consequence of 7.3 and 10.2 there is the following numerical characterization of homogeneous complete intersections of codimension 0:

11.2 Theorem *Let $Q \to A$ be a noetherian specialization and S as before. Then*

$$C := A \otimes_Q S = A[T]/(F_0, \ldots, F_n)A[T]$$

is a finite complete intersection over A if and only if the elimination ideal $\mathfrak{T}_0(S)A$ (corresponding to the representation of C specified) is the unit ideal.

Later we shall see that the generic elimination ideal $\mathfrak{T}_0(S)$ in 11.1 is a principal ideal, cf. 12.4 and, for a special case, 11.8 at the end of the current section. Here we are going to study some simple geometric properties of $\mathrm{Proj}\, S$.

Until the end of this section we will assume without further comment that the homogeneous generic polynomials $F_0, \ldots, F_n$ are non-zero and, moreover, constitute an admissible sequence.

The Lasker-Noether decomposition of the ideal $(F_0, \ldots, F_n) \subseteq Q[T]$, which we have already described in the more general set-up of Section 8, has only components which are homogeneous of height $n+1$. Thus we have the decomposition

$$(F_0, \ldots, F_n) = \mathfrak{Q}^* \cap \mathfrak{T}^*$$

where $\mathfrak{Q}^*$ is the primary component belonging to the irrelevant (prime) ideal $Q[T]_+ = (T_0, \ldots, T_n)$ and $\mathfrak{T}^*$ is the intersection of the remaining components. In the residue class ring $S = Q[T]/(F_0, \ldots, F_n)$ there is the corresponding decomposition

$$0 = \mathfrak{Q} \cap \mathfrak{T}$$

with $\mathfrak{Q} := \mathfrak{Q}^*/(F_0, \ldots, F_n)$ and the ideal $\mathfrak{T} := \mathfrak{T}^*/(F_0, \ldots, F_n) = \mathfrak{T}(S)$ of inertia forms, cf. Section 10.

The following simple statements about these ideals will be used frequently.

11.3 $\mathfrak{Q}$ *is nothing else but the Q-torsion of S.*

Proof. $\mathfrak{Q}$ is the kernel of the canonical homomorphism $S \to S_{\mathfrak{M}}$ corresponding to the prime ideal $\mathfrak{M} := S_+$. Obviously, the Q-torsion of S belongs to the kernel. Conversely, if $fg = 0$ in S and f homogeneous, $g \notin \mathfrak{M}$, then f is annihilated by the constant term $g_0 \neq 0$ of g.   □

11.4 $\mathfrak{Q}$ *contains the ideal* $S_{(\sigma+1)} = \bigoplus_{m > \sigma} S_m$, *where σ is the number*

$$\sigma := \sum_{j=0}^{n} \delta_j - \sum_{i=0}^{n} \gamma_i \,.$$

σ *is non-negative.*

Proof. $\mathfrak{Q}$ is the kernel of the homomorphism $S \to L \otimes_Q S = S_{(L)}$, L being the quotient field of Q, as was proved just above. Thus one has to show that all homogeneous components of degree $> \sigma$ of $S_{(L)}$ vanish. But by 7.2 the L-algebra $S_{(L)}$ is a finite complete intersection with Poincaré series

$$\frac{(1 - Z^{\delta_0}) \cdots (1 - Z^{\delta_n})}{(1 - Z^{\gamma_0}) \cdots (1 - Z^{\gamma_n})}$$

which is a polynomial of degree $\sigma (\geq 0)$.   □

The non-negative number $\sigma = \sum_j \delta_j - \sum_i \gamma_i$ occurring in 11.4 is crucial in the whole set-up. In general $\mathfrak{Q}$ differs from $S_{(\sigma+1)}$, cf. Supplement 5.

11.5 *The ideal $\mathfrak{T} = \mathfrak{T}(S)$ of inertia forms is the S-annihilator of $\mathfrak{Q}$ and of $S_{(\sigma+1)}$:*

$$\mathfrak{T} = \mathrm{Ann}_S \mathfrak{Q} = \mathrm{Ann}_S S_{(\sigma+1)} \,.$$

Proof. The first equation has already been established in Section 10 (before 10.3), and by the very definition of $\mathfrak{T}$ the ideal $\mathrm{Ann}_S S_{(\sigma+1)}$ belongs to $\mathfrak{T}$. $\square$

Among the primary components of $\mathfrak{T}^\star \subseteq Q[T]$ there is the distinguished one called **p r i n c i p a l c o m p o n e n t** or **t o r i c c o m p o n e n t** and denoted by $\mathfrak{H}^\star$; its image in S is $\mathfrak{H} := \mathfrak{H}^\star/(F_0, \ldots, F_n)$. It can be defined by

$$\mathfrak{H}^\star := (F_0, \ldots, F_n)Q[T, T^{-1}] \cap Q[T] = \mathfrak{T}^\star Q[T, T^{-1}] \cap Q[T]$$

using the ring $Q[T, T^{-1}] = Q[T_0, \ldots, T_n, T_0^{-1}, \ldots, T_n^{-1}] = Q[T]_{T_0 \cdots T_n}$ of Laurent polynomials, see Section 8. The ideal $\mathfrak{H}^\star$ is the intersection of those primary components of $(F_0, \ldots, F_n)$ the zero sets of which meet the (weighted) torus $\mathrm{D}_+(T_0 \cdots T_n) \subseteq \mathrm{Proj}\, Q[T] = \mathbb{P}_\gamma^n(Q)$. As we have seen in Section 8:

$\mathfrak{H}^\star$ *is prime and therefore a prime primary component of* $\mathfrak{T}^\star$.

Besides $\mathfrak{H}^\star$ there may be other primary components of $\mathfrak{T}^\star$ which we will call the **p e r i p h e r a l c o m p o n e n t s**. A peripheral component is characterized by the property that its set of zeros is contained in one of the hyperplanes $\mathrm{V}_+(T_i) \subseteq \mathbb{P}_\gamma^n(Q)$, $i = 0, \ldots, n$. The peripheral components contained in the hyperplane $\mathrm{V}_+(T_i)$ are in one-to-one correspondence to the relevant associated prime ideals of height n of the ideal (of height $\geq n$) generated by the polynomials $F_j|_{T_i=0}$, $j = 0, \ldots n$, in $Q[T_0, \ldots, \hat{T}_i, \ldots, T_n]$.

The case $\mathfrak{H}^\star = \mathfrak{T}^\star$ is described in a combinatorial way in Lemma 8.8: It is the strictly admissible case.

In the saturated classical case $\gamma_0 = \cdots = \gamma_n = 1$ even the monomials T^ν, $\nu \in N_j = \mathrm{N}_{\delta_j}(\gamma)$, of a single polynomial F_j generate the ideal $(T_0, \ldots, T_n)^{\delta_j}$ of codimension $n + 1$. Lemma 8.8 and Lemma 8.6 therefore yield the following well-known classical result, cf. [19], for example:

11.6 Theorem *In the saturated classical case* $\gamma_0 = \cdots = \gamma_n = 1$ *the ideals* $(F_0, \ldots, F_r)$ *are prime ideals of codimension* $r + 1$ *for* $r < n$, *and*

$$(F_0, \ldots, F_n) = \mathfrak{Q}^\star \cap \mathfrak{H}^\star$$

is the intersection of the primary ideal $\mathfrak{Q}^\star$ *belonging to the irrelevant ideal and the principal prime component* $\mathfrak{H}^\star$ *which coincides with the ideal of inertia* $\mathfrak{T}^\star$ *of* $(F_0, \ldots, F_n)$.

More generally, Lemma 8.8 implies, together with 8.6:

11.7 *In the strictly admissible case every ideal* $\mathfrak{F}_J := \sum_{j \in J} Q[T] F_j$ *is a prime ideal of codimension* $|J|$ *for every proper subset* $J \subset \{0, \ldots, n\}$.

As a final application of Lemma 8.8 we prove that in the strictly admissible case the general elimination ideal $\mathfrak{T}_0(S) = \mathfrak{T}_0^\star = \mathfrak{T}^\star \cap Q$ is a principal prime ideal. Its generator is a prime polynomial, uniquely determined up to sign, which might serve as the resultant of $F_0, \ldots, F_n$. This approach will be discussed in detail along with the treatment of the general situation in Sections 14 and 15.

We will have constant use of the following concepts already introduced in Section 9 (in more general circumstances).

Definition For $m \in \mathbb{N}$ and every $N \subseteq N_m(\gamma)$ the group $\Gamma(N)$ is the subgroup

$$\Gamma(N) := \sum_{\nu, \mu \in N} \mathbb{Z} (\nu - \mu)$$

of $\mathbb{Z}^{n+1}$. This is even a subgroup of $\Gamma = \mathrm{Syz}(\gamma) := \{\alpha : \langle \alpha, \gamma \rangle = 0\} \subset \mathbb{Z}^{n+1}$ of syzygies of the weights γ. For $N = N_m(\gamma)$ we set

$$\Gamma(m) = \Gamma(m; \gamma) := \Gamma(N_m(\gamma)).$$

Notice that in case $N \neq \emptyset$ the group $\Gamma(N)$ is already generated by the differences $\nu - \mu_0$, $\nu \in N$, for a fixed $\mu_0 \in N$.

For the generic polynomials $F_j = \sum_{\nu \in N_j} U_{j\nu} T^\nu$, $j = 0, \ldots, n$, we will usually abbreviate $\Gamma(N_j)$ by

$$\Gamma_j := \Gamma(N_j), \quad j = 0, \ldots, n.$$

The part $\mathfrak{H}_0^\star = \mathfrak{H}^\star \cap Q$ of degree zero in the principal component $\mathfrak{H}^\star$ is the kernel of the natural homomorphism

$$Q \longrightarrow H := Q[T, T^{-1}]/(\mathfrak{H}^\star) = Q[T, T^{-1}]/(F_0, \ldots, F_n).$$

H is a graded algebra $\bigoplus_{m \in \mathbb{Z}} H_m = H_0[Z_0, Z_0^{-1}]$, where Z_0 is a Laurent monomial $Z_0 := T_0^{\beta_0} \cdots T_n^{\beta_n}$ with $\beta_i \in \mathbb{Z}$ and $\sum_i \beta_i \gamma_i = \gcd(\gamma_0, \ldots, \gamma_n)$. The image of Q lies in H_0. As in Section 9 this image can be identified with the ring

$$Q^\flat[G_0, \ldots, G_n] \subseteq H_0 \subseteq H = Q^\flat[T, T^{-1}]$$

where

$$Q^\flat := \mathbb{Z}[U_{j\nu} : j = 0, \ldots, n, \ \nu \in N'_j := N_j \smallsetminus \{\nu(j)\}],$$

$\nu(j) \in N_j$ chosen arbitrarily, and

$$G_j := U_{j\nu(j)} - T^{-\nu(j)} F_j = -\sum_{\nu \in N'_j} U_{j\nu} T^{\nu - \nu(j)}.$$

Furthermore, for every $\mathbb{Z}$-base $\alpha(1), \ldots, \alpha(n)$ of $\mathrm{Syz}(\gamma)$ one has

$$H_0 = Q^\flat[T^{\alpha(1)}, \ldots, T^{\alpha(n)}].$$

Because of $\mathfrak{H}_0^\star \cap \mathbb{Z} = 0$ (see Supplement 1) the codimension of $\mathfrak{H}_0^\star$ in Q, which is the codimension (≥ 1) of the projection of the principal component $V_+(\mathfrak{H}^\star)$ in $\mathrm{Spec}\, Q$, may be computed over the rationals as the codimension of $\mathfrak{H}_0^\star Q_{(\mathbb{Q})}$ in $Q_{(\mathbb{Q})}$. This codimension can be > 1, cf. Supplement 2. In the strictly admissible case, however, the codimension is 1.

11.8 Theorem *In the strictly admissible case the generic elimination ideal* $\mathfrak{T}_0(S)$, $S = Q[T]/(F_0, \ldots, F_n)$, *is a principal prime ideal* $\neq 0$ *in* $Q = \mathbb{Z}[U_{j\nu}]_{j,\nu}$.

Proof. Since Q is a unique factorization domain and $\mathfrak{T}_0^\star = \mathfrak{T}_0(S)$ is prime and $\neq 0$ we will just have to show that the codimension of $\mathfrak{T}_0^\star$ is ≤ 1. But this is true by 9.1. $\qquad\square$

There is an explicit combinatorial characterization of the case that $\mathfrak{H}_0^\star = \mathfrak{H}^\star \cap Q$ is of codimension 1 in Q (and hence a principal prime ideal $\neq 0$). This condition means that the extension $Q^\flat[G_0, \ldots, G_n] \subseteq H_0$ is algebraic or that the sequence $G_0, \ldots, G_n$ contains n elements which are algebraically independent over $Q^\flat$. As a corollary of 9.2 we get:

11.9 Theorem *The following conditions are equivalent*:
(1) $\mathfrak{H}_0^\star$ *is a principal prime ideal* $\neq 0$ *in* Q.
($1'$) *The quotient field of* H_0 *is finite over the quotient field of the image* $Q^\flat[G_0, \ldots, G_n] = Q/\mathfrak{H}_0^\star$ *of* Q *in* H_0.
(2) *There is a proper subset* J' *of* $\{0, \ldots, n\}$ *with* $|J'| = n$ *such that for every* $J \subseteq J'$ *the group* $\Gamma_J = \sum_{j \in J} \Gamma_j$ *has rank* $\geq |J|$.
For a set J' *as in condition* (2) *the degree of the quotient field of* H_0 *over the quotient field of* $Q^\flat[G_j : j \in J']$ *is divisible by the group index* $[\Gamma : \Gamma_{J'}]$.

The additional remark in Theorem 11.9 follows simply from the inclusions $Q^\flat[G_j : j \in J'] \subseteq Q^\flat[T^\alpha : \alpha \in \Gamma_{J'}] \subseteq H_0$.

Condition (2) is satisfied for every subset $J' \subset \{0, \ldots, n\}$ with $|J'| = n$ if the sequence $F_0, \ldots, F_n$ is strictly admissible, cf. Lemma 9.6. In this important situation we have the following picture:

11.10 Theorem *Let the sequence* $F_0, \ldots, F_n$ *of homogeneous generic polynomials be strictly admissible. Let* L *denote the quotient field of* H_0 *and* $K \subseteq L$ *the quotient field of the image* $Q^\flat[G_0, \ldots, G_n] = Q/\mathfrak{H}_0^\star$ *of* Q *in* H_0. *Then the following holds*:

(1) *For every* $i \in \{0, \ldots, n\}$ *the elements* G_j, $j \neq i$, *are algebraically independent over* $Q^\flat$, *and the degree of* L *over the quotient field* K_i *of* $Q^\flat[G_j : j \neq i]$ *is divisible by the group index* $[\Gamma : \sum_{j \neq i} \Gamma_j]$. *Furthermore*,

$$[L : K_i] = \gcd(\gamma_0, \ldots, \gamma_n)(\textstyle\prod_{j \neq i} \delta_j)/\gamma_0 \cdots \gamma_n.$$

(2) K *is nothing else but the quotient field of* $Q^\flat[T^{\beta(1)}, \ldots, T^{\beta(n)}]$, *where* $\beta(1), \ldots, \beta(n)$ *is a* $\mathbb{Z}$-*base of* $\Gamma_0 + \cdots + \Gamma_n$. *In particular*,

$$[L : K] = [\Gamma : (\Gamma_0 + \cdots + \Gamma_n)].$$

Proof. (1) Only the formula for the degree has yet to be proved. For this we may assume $\gcd(\gamma_0, \ldots, \gamma_n) = 1$ and $i = 0$.

Let $Q' := \mathbb{Z}[U_{j\nu} : j \neq 0] \subseteq Q$, $H' := Q'[T, T^{-1}]/(F_1, \ldots, F_n)$ and $Q'^{\triangleright} := Q' \cap Q^{\triangleright}$. Then, by Supplement 7, $H'_0 = Q'^{\triangleright}[T^{\alpha(1)}, \ldots, T^{\alpha(n)}]$ has rank $\delta_1 \cdots \delta_n / \gamma_0 \gamma_1 \cdots \gamma_n$ over the image $Q'^{\triangleright}[G_1, \ldots, G_n]$ of Q' in H'_0. This is also the rank of $H_0 = H'_0[U_{0\nu} : \nu \in N'_0]$ over $Q^{\triangleright}[G_1, \ldots, G_n] = Q'^{\triangleright}[U_{0\nu} : \nu \in N'_0][G_1, \ldots, G_n]$.

(2) is an immediate consequence of Lemma 9.2.   ☐

Consider a sequence $F_0, \ldots, F_n$ of homogeneous generic polynomials, admissible as throughout this section, with principal component $\mathfrak{H}^{\star}$ and its irreducible variety $V_+(\mathfrak{H}^{\star})$, which is $\operatorname{Proj}(S/\mathfrak{H}) = \operatorname{Proj}(Q[T]/\mathfrak{H}^{\star}) \subseteq \mathbb{P}^n_{\gamma}(Q)$. The image of the set $V_+(\mathfrak{H}^{\star})$ in $\operatorname{Spec} Q$ is the irreducible variety $V(\mathfrak{h}) = \operatorname{Spec}(Q/\mathfrak{h})$, $\mathfrak{h} := \mathfrak{H}^{\star}_0 = \mathfrak{H}_0$. The morphism of elimination

$$V_+(\mathfrak{H}^{\star}) \longrightarrow V(\mathfrak{h})$$

for the principal (toric) component is generically finite (by definition) if and only if the quotient field of H_0 is algebraic over $Q/\mathfrak{h}$, which is characterized in Theorem 11.9. In this case the degree $[L : K]$ of the quotient fields L, K of H_0 and $Q/\mathfrak{h}$, respectively, is called the p r i n c i p a l d e g r e e o f e l i m i n a t i o n.

The concept of the (total) d e g r e e o f e l i m i n a t i o n will be introduced in Section 14. The definition implies directly that for strictly admissible sequences of homogeneous generic polynomials $F_0, \ldots, F_n$ the principal degree of elimination coincides with the total degree. Therefore we refer to it simply as the degree of elimination. By Theorem 11.10 it can be computed as the group index $[\Gamma : (\Gamma_0 + \cdots + \Gamma_n)]$.

The degree of elimination may be > 1, even in the saturated case, i. e. for strictly admissible pairs γ, δ, if the dimension is large enough: $n \geq 4$, cf. Supplement 10. On the other hand, the degree of elimination is 1 for all strictly admissible pairs with $n \leq 3$, cf. Supplement 11.

Supplements

Throughout the supplements we refer to the main case of elimination for homogeneous generic polynomials under admissibility conditions.

1. The intersections $\mathfrak{H}^{\star} \cap \mathbb{Z}$ and $\mathfrak{T}^{\star} \cap \mathbb{Z} (\subseteq \mathfrak{H}^{\star} \cap \mathbb{Z})$ are zero. ($k \otimes_{\mathbb{Z}} H \neq 0$ for all commutative rings $k \neq 0$.)

2. Assume that there is a proper subset J of $\{0, \ldots, n\}$ such that F_j, $j \in J$, generate a prime ideal in $Q[T]$. Then $\operatorname{codim} \mathfrak{H}^{\star}_0 \leq n + 1 - |J|$. (One proceeds as in the proof of 11.8.) In particular

$$1 \leq \operatorname{codim} \mathfrak{H}^{\star}_0 \leq n + 1.$$

All the codimensions $s + 1$ between 1 and $n + 1$ do occur. (Example: $F_0, \ldots, F_s$ have only one term each, and $(F_{s+1}, \ldots, F_n)$ is a prime ideal.)

In general, the equation $\operatorname{codim} \mathfrak{H}_0^* = s + 1$ does not imply that there is a subset $J \subset \{0, \ldots, n\}$ with $|J| = n - s$ such that F_j, $j \in J$, generate a prime ideal. (Example: $n = 1$, $\gamma = (2, 3)$, $\delta = (8, 9)$.)

3. Let $n = 1$. The pair $\gamma = (\gamma_0, \gamma_1)$, $\delta = (\delta_0, \delta_1)$ is strictly admissible if and only if δ_0 and δ_1 are common multiples of γ_0 and γ_1. In this (saturated) case a generator of the elimination ideal (cf. 11.8) can be constructed using a Sylvester determinant.

Let us consider the classical case $\gamma_0 = \gamma_1 = 1$, first. In this case the elimination ideal is the intersection

$$Q \cap (Q[X]f_0 + Q[X]f_1),$$

$f_0 = F_0(1, X) = U_0 + U_1 X + \cdots + U_r X^r$, $f_1 = F_1(1, X) = V_0 + V_1 X + \cdots + V_s X^s$, $r := \delta_0$, $s := \delta_1$. By Cramer's rule the Sylvester determinant

$$\operatorname{Sylv}(f_0, f_1) = \begin{vmatrix} U_0 & \cdots & \cdots & U_r & & & \\ & \ddots & & & \ddots & & \\ & & U_0 & \cdots & \cdots & U_r \\ V_0 & \cdots & \cdots & V_s & & \\ & \ddots & & & \ddots & \\ & & V_0 & \cdots & \cdots & V_s \end{vmatrix} \in Q$$

of degree s in the U's and degree r in the V's multiplies the free Q-module with basis $1, X, \ldots, X^{r+s-1}$ onto the module with basis $f_0, X f_0, \ldots, X^{s-1} f_0, f_1, X f_1, \ldots, X^{r-1} f_1$. Thus $\operatorname{Sylv}(f_0, f_1)$ belongs to the elimination ideal. Since it is a prime polynomial, it is a generator.

In the general case we may assume $\gcd(\gamma_0, \gamma_1) = 1$. Any monomial $T_0^{\nu_0} T_1^{\nu_1}$ of γ-degree $\delta_j = r_j \gamma_0 \gamma_1$ is of the form $(T_0^{\gamma_1})^{\mu_0} (T_1^{\gamma_0})^{\mu_1}$, $\mu_0 + \mu_1 = r_j$, $j = 0, 1$. Therefore $F_j = G_j(T_0^{\gamma_1}, T_1^{\gamma_0})$, where

$$G_j = \sum\nolimits_{\mu_0 + \mu_1 = r_j} U_{j\mu_0\mu_1} X_0^{\mu_0} X_1^{\mu_1}$$

are the homogeneous generic polynomials of degree r_j of the classical case. Then the Sylvester determinant $\operatorname{Sylv}(G_0, G_1)$ generates the elimination ideal of (F_0, F_1).

4. For $F_j = \sum_{i=0}^n U_{ji} T_i$, $j = 0, \ldots, n$, with $\gamma_0 = \cdots = \gamma_n = 1$ the elimination ideal is generated by the determinant $\det (U_{ij})_{0 \le i, j \le n}$.

5. The irrelevant component $\mathfrak{Q}$ of $S = Q[T]/(F_0, \ldots, F_n)$ may be strictly larger than $S_{(\sigma+1)}$, i.e. the homogeneous part S_m may have non-trivial Q-torsion for $m \le \sigma = \sum_i (\delta_i - \gamma_i)$. For example, the Q-rank of S_m (as given by the coefficient of Z^m in the Poincaré series $\prod_i (1 - Z^{\delta_i})(1 - Z^{\gamma_i})^{-1}$) may be zero while the minimal number of Q-generators of S_m (as given by the coefficient of Z^m in the Poincaré series $\prod_i (1 - Z^{\gamma_i})^{-1}$) is not zero. (A good candidate for m is $\sigma - 1$.)

6. a) For $\gamma = (2, 3, 5)$ and $\delta = (4, 5, 6)$ the elimination ideal in the saturated case is generated by $U_{0;200} U_{1;001}^2 U_{2;020}$ (and especially not reduced).
b) For $\gamma = (2, 3, 4)$ and $\delta = (4, 5, 6)$ the elimination ideal is generated by

$$U_{0;001} U_{1;110} U_{2;020} (U_{0;200} U_{2;101} - U_{0;001} U_{2;300}).$$

The projection of the peripheral component with radical $(U_{0;001}, U_{2;020}, T_0)$ is contained in the projection of the other peripheral components.

7. Assume that $\gcd(\gamma_0, \ldots, \gamma_n) = 1$. If the n homogeneous generic polynomials $F_1, \ldots, F_n$ generate a prime ideal (of codimension n), the sequence of coefficients of the Poincaré series
$$\mathcal{P} = \frac{(1 - Z^{\delta_1}) \cdots (1 - Z^{\delta_n})}{(1 - Z^{\gamma_0})(1 - Z^{\gamma_1}) \cdots (1 - Z^{\gamma_n})}$$
of the algebra $Q[T]/(F_1, \ldots, F_n)$ gets stationary with limit $\delta_1 \cdots \delta_n / \gamma_0 \gamma_1 \cdots \gamma_n$. (The product $(1 - Z)\mathcal{P}$ is a polynomial because the prime factors of the denominators are cancelled completely, cf. Suppl. 1 of Sect. 8; note that one can switch to the saturated case.)

8. Let $\gamma = (\gamma_0, \ldots, \gamma_n)$ and $\delta = (\delta_0, \ldots, \delta_r)$ consist of positive weights and degrees, respectively. Let $g := \gcd(\gamma_0, \ldots, \gamma_n)$ and $\gamma_i' := \gamma_i/g$, $\delta_j' := \delta_j/g$ for $i = 0, \ldots, n, j = 0, \ldots, r$. Then the pair γ, δ is admissible (or strictly admissible) if and only if g divides all δ_j, i.e. $\delta_0', \ldots, \delta_r'$ are integers, and the pair γ', δ' is admissible (strictly admissible, resp.).

If one of these conditions is fulfilled, then $\mathrm{Syz}(\gamma) = \mathrm{Syz}(\gamma')$ and $\Gamma(\delta_j, \gamma) = \Gamma(\delta_j', \gamma')$, hence
$$[\mathrm{Syz}(\gamma) : \Gamma_{\delta_0}(\gamma) + \cdots + \Gamma_{\delta_r}(\gamma)] = [\mathrm{Syz}(\gamma') : \Gamma_{\delta_0'}(\gamma') + \cdots + \Gamma_{\delta_r'}(\gamma')].$$
($N_{\delta_j}(\gamma) = N_{\delta_j'}(\gamma')$ holds for every j.)

In particular, to compute the degree of elimination in the strictly admissible case, one can switch easily to the case that the weights $\gamma_0, \ldots, \gamma_n$ are coprime.

9. Let the pair $\gamma = (\gamma_0, \ldots, \gamma_n)$, $\delta = (\delta_0, \ldots, \delta_n)$ be strictly admissible. If $\mathrm{Mon}(\gamma)$ is a complete intersection, then the degree of elimination $[\Gamma : (\Gamma_0 + \cdots + \Gamma_n)]$ is 1. (4.6. Especially, the degree of elimination is 1, if one of the weights is 1.)

10. a) The pair $\gamma := (6, 6, 10, 10, 15, 11)$, $\delta := (30, 30, 30, 30, 22, 22)$ is strictly admissible. The degree of elimination $[\Gamma : (\Gamma_0 + \cdots + \Gamma_5)]$ is 2. (The last component of every $\alpha \in N_{30}(\gamma)$ and every $\alpha \in N_{22}(\gamma)$ is 0 or 2.) Instead of $\gamma_5 = 11$ and $\delta_4 = \delta_5 = 22$ one can take also $\gamma_5 = 13$ and $\delta_4 = \delta_5 = 26$ or $\gamma_5 = 17$ and $\delta_4 = \delta_5 = 34$.

For the strictly admissible pair $(12, 12, 15, 15, 20, 19), (60, 60, 60, 60, 57, 57)$ the degree of elimination is 3. (For other examples cf. Sect. 14, Suppl. 9.)

b) Each of the pairs $\gamma := (6, 10, 14, 105, \gamma_4)$, $\delta := (210, 210, 210, 2\gamma_4, 2\gamma_4)$ is strictly admissible for $\gamma_4 \in \{97, 101, 103, 107, 109, 113\}$. The degree of elimination $[\Gamma : (\Gamma_0 + \cdots + \Gamma_4)]$ is 2 in all these cases.

Each of the pairs $\gamma := (12, 15, 21, 140, \gamma_4)$, $\delta := (420, 420, 420, 3\gamma_4, 3\gamma_4)$ is strictly admissible with 3 as degree of elimination for $\gamma_4 \in \{131, 137, 143, 149\}$.

(For $n \leq 3$, strictly admissible pairs $\gamma = (\gamma_0, \ldots, \gamma_n)$, $\delta := (\delta_0, \ldots, \delta_n)$ have always degree of elimination 1, see the next supplement.)

11. In this supplement we prove the following

Theorem: *In case $n \leq 3$, for all strictly admissible pairs $\gamma = (\gamma_0, \ldots, \gamma_n)$, $\delta = (\delta_0, \ldots, \delta_n)$ the degree of elimination is 1.*

It suffices to consider the case $\gcd(\gamma_0, \ldots, \gamma_n) = 1$, cf. Suppl. 8. In addition, we assume $n \geq 1$.

First we introduce some notations. Let

$$\Gamma((d_0,\ldots,d_r);\gamma) := \sum_{j=0}^{r}\Gamma(d_j;\gamma) \subseteq \Gamma := \mathrm{Syz}(\gamma) \subseteq \mathbb{Z}^{n+1}$$

for an $(r+1)$-tuple $d = (d_0,\ldots,d_r) \in (\mathbb{N}^*)^{r+1}$. For any subsequence $\hat{\gamma} = (\gamma_{i_1},\ldots,\gamma_{i_s})$ of γ we identify $\mathrm{Syz}(\hat{\gamma})$ with a subgroup of $\mathrm{Syz}(\gamma)$ in the canonical way; then $\Gamma(d;\hat{\gamma}) \subseteq \Gamma(d;\gamma)$. In these notations we do not assume that $\gcd(\gamma_{i_1},\ldots,\gamma_{i_s}) = 1$. For arbitrary $a,b \in \mathbb{N}^*$ let $v(a,b)$ be 1 if a is a multiple of b and 0 otherwise. Then the $(r+1) \times (n+1)$-matrix

$$\mathfrak{V} = \mathfrak{V}_\gamma(d) := (v(d_j,\gamma_i))_{0\leq j\leq r,0\leq i\leq n}$$

is called the *V-matrix* of d (with respect to γ). By definition, the graph $G_\gamma(d)$ on the set $\{0,\ldots,n\}$ as vertices contains the (undirected) edge connecting i_1,i_2 if there is a j such that $v(d_j,\gamma_{i_1}) = v(d_j,\gamma_{i_2}) = 1$, i.e. if $\mathfrak{V}(d)$ contains a row with 1 in the two (different) columns i_1,i_2. Furthermore, it will be convenient to set $(i_1\ldots i_s) := \gcd(\hat{\gamma})$ for a subsequence $\hat{\gamma} = (\gamma_{i_1},\ldots,\gamma_{i_s})$ of γ.

a) $\Gamma(d_0;\gamma) \subseteq \Gamma(td_0;\gamma)$ for all $d_0,t \in \mathbb{N}^*$. (Let $\mu,\nu \in \mathrm{N}_{d_0}(\gamma)$. Then $\mu - \nu = t\mu - ((t-1)\mu + \nu) \in \Gamma(td_0;\gamma)$.)

b) (*Reduction in codimension* 1) Let $t \in \mathbb{N}^*$ be a common divisor of $\gamma_1,\ldots,\gamma_n$ and $\gamma_0' := \gamma_0$, $\gamma_i' := \gamma_i/t$ for $i = 1,\ldots,n$. As at the end of Sect. 3 we will use the canonical reduction map $R : \mathbb{Z} \to \mathbb{Z}$. The canonical isomorphism $\mathrm{Syz}(\gamma) \to \mathrm{Syz}(\gamma')$ induces an isomorphism $\Gamma(d;\gamma) \to \Gamma(R(d);\gamma')$ as well as an isomorphism $\mathrm{Syz}(\gamma)/\Gamma(d;\gamma) \to \mathrm{Syz}(\gamma')/\Gamma(R(d);\gamma')$.

Let γ,δ be strictly admissible. Then all δ_j are divisible by t (since T_0 is not a factor of F_j), hence $R(\delta) = \delta/t$. Since all the canonical mappings $\mathrm{N}_{\delta_j}(\gamma) \to \mathrm{N}_{\delta_j/t}(\gamma')$ are bijective, the pair $\gamma',\delta/t$ is strictly admissible with the same degree of elimination. (By the way, the pair $\gamma, t\delta'$ is strictly admissible if γ', δ' is a strictly admissible pair.)

As a consequence, in order to prove the theorem one can assume that γ is *reduced*, as we will say, i.e. that any n of the γ_i are relatively prime.

c) If $v(d_0,\gamma_{i_1}) = v(d_0,\gamma_{i_2}) = 1$ for $i_1 \neq i_2$, then $\Gamma(d_0;\gamma)$ contains the syzygy

$$(-\gamma_{i_2}/(i_1 i_2),\gamma_{i_1}/(i_1 i_2)).$$

$((-\gamma_{i_2}/(i_1 i_2),\gamma_{i_1}/(i_1 i_2)) \in \Gamma(\mathrm{lcm}(\gamma_{i_1},\gamma_{i_2});(\gamma_{i_1},\gamma_{i_2})) \subseteq \Gamma(d_0;(\gamma_{i_1},\gamma_{i_2}))$ by a), and $\Gamma(d_0;(\gamma_{i_1},\gamma_{i_2})) \subseteq \Gamma(d_0;\gamma)$.)

A consequence is: If $v(d_0,\gamma_i) = 1$ for all $i = 0,\ldots,n$, then $\Gamma(d_0;\gamma) = \Gamma$. In other words: If a V-matrix $\mathfrak{V}_\gamma(d)$ contains a row $(1,\ldots,1)$, then $\Gamma(d;\gamma) = \Gamma$.

d) If γ,δ is a strictly admissible pair, then every column of $\mathfrak{V}_\gamma(\delta)$ contains at least two 1's. (For every i_0 the ideal $(T_i : i \neq i_0)$ contains at most $n-1$ of the polynomials $F_0,\ldots,F_n$.)

e) The theorem holds for $n = 1$ and $n = 2$. (By d), for $n = 1$ the V-matrix is

$$\begin{pmatrix} 1 & 1 \\ 1 & 1 \end{pmatrix}$$

and for $n = 2$ the matrix $\mathfrak{V}_\gamma(\delta)$ contains the row $(1,1,1)$ or coincides with

$$\begin{pmatrix} 0 & 1 & 1 \\ 1 & 0 & 1 \\ 1 & 1 & 0 \end{pmatrix}$$

after a suitable rearranging of the δ's. Now use c).)

f) Let $n = 2$. If the V-matrix $\mathfrak{V}_\gamma(d_0, d_1)$ is of type

$$\begin{pmatrix} 1 & 1 & \cdot \\ \cdot & 1 & 1 \end{pmatrix}$$

then $\Gamma' := \Gamma((d_0, d_1); (\gamma_0, \gamma_1, \gamma_2)) = \Gamma$. (One may assume $d_0 = \mathrm{lcm}(\gamma_0, \gamma_1)$, $d_1 = \mathrm{lcm}(\gamma_1, \gamma_2)$. Furthermore, by b) one can assume that γ is reduced. Then $d_0 = \gamma_0\gamma_1$, $d_1 = \gamma_1\gamma_2$, and Γ' contains $(-\gamma_1, \gamma_0, 0)$, $(0, -\gamma_2, \gamma_1)$. If $d_1 - \gamma_0 \in \mathrm{Mon}(\gamma_1, \gamma_2)$, then Γ' contains a syzygy of type $(1, \cdot, \cdot)$, hence $\Gamma' = \Gamma$. Similarly, if $d_0 - \gamma_2 \in \mathrm{Mon}(\gamma_0, \gamma_1)$, then $\Gamma' = \Gamma$. One of these cases occurs. Otherwise, $d_1 - \gamma_0 = \gamma_1\gamma_2 - \gamma_0 \le g_{(\gamma_1, \gamma_2)} = \gamma_1\gamma_2 - \gamma_1 - \gamma_2$, hence $\gamma_0 \ge \gamma_1 + \gamma_2$. Similarly, $\gamma_2 \ge \gamma_0 + \gamma_1$. Contradiction.)

g) Let $n = 3$. If the V-matrix $\mathfrak{V}_\gamma(d_0, d_1)$ is of type

$$\begin{pmatrix} 1 & 1 & 1 & \cdot \\ \cdot & \cdot & 1 & 1 \end{pmatrix}$$

then $\Gamma' := \Gamma((d_0, d_1); \gamma) = \Gamma$. (By reduction one may assume $(023) = (123) = 1$. Then
$$\mathrm{Syz}(\gamma_0, \gamma_1, \gamma_2) + \mathrm{Syz}(\gamma_0, \gamma_2, \gamma_3) + \mathrm{Syz}(\gamma_1, \gamma_2, \gamma_3) \subseteq \Gamma'$$
by c) and f). But there are syzygies $(\cdot, 0, \cdot, (02))$ in $\mathrm{Syz}(\gamma_0, \gamma_2, \gamma_3)$ and $(0, \cdot, \cdot, (12))$ in $\mathrm{Syz}(\gamma_1, \gamma_2, \gamma_3)$, hence $(\cdot, \cdot, \cdot, (012)) \in \Gamma'$.

Later the statement will only be used in case γ is already reduced.)

h) Let $n = 3$. If $\{3\}$ is a connected component of $G_\gamma(\delta)$ then

$$\mathfrak{V}_\gamma(\delta) = \begin{pmatrix} 1 & 1 & 1 & 0 \\ 1 & 1 & 1 & 0 \\ 0 & 0 & 0 & 1 \\ 0 & 0 & 0 & 1 \end{pmatrix}$$

after a suitable renumbering of the δ_j's. (Use part d).) Therefore, to prove the theorem for $n = 3$, one has to consider three cases:

Case I : $G_\gamma(\delta)$ is connected.
Case II : The connected components of $G_\gamma(\delta)$ are $\{0, 1\}$, $\{2, 3\}$.
Case III: The connected components of $G_\gamma(\delta)$ are $\{0, 1, 2\}$, $\{3\}$.

From now only the case $n = 3$ with a reduced γ will be considered. We define $\Gamma' := \Gamma(\delta; \gamma)$.

i) Case I: Let $G_\gamma(\delta)$ be connected. Three subcases can occur:
(1) $\mathfrak{V}_\gamma(\delta)$ contains a row $(1, 1, 1, 1)$. Then $\Gamma' = \Gamma$ by c).
(2) $\mathfrak{V}_\gamma(\delta)$ contains a row with three 1's, say $\mathfrak{V}_\gamma(\delta_0) = (1, 1, 1, 0)$. Because $G_\gamma(\delta)$ is connected, there is another row, say $\mathfrak{V}_\gamma(\delta_1)$, of type $(\cdot, \cdot, \cdot, 1)$ which connects 3 with some point of $\{0, 1, 2\}$. We may assume $\mathfrak{V}_\gamma(\delta_1) = (\cdot, \cdot, 1, 1)$. Now $\Gamma' = \Gamma$ by g).
(3)   Every row contains at most two 1's. By d) every row and every column of $\mathfrak{V}_\gamma(\delta)$ contains exactly two 1's. Because of connectedness the matrix

$$\begin{pmatrix} 1 & 1 \\ 1 & 1 \end{pmatrix}$$

is not a submatrix of $\mathfrak{V}_\gamma(\delta)$. Thus there are two rows, say $\mathfrak{V}_\gamma(\delta_0)$, $\mathfrak{V}_\gamma(\delta_1)$, of type $(\cdot, \cdot, \cdot, 0)$. Then $\Gamma((\delta_0, \delta_1); (\gamma_0, \gamma_1, \gamma_2)) = \mathrm{Syz}(\gamma_0, \gamma_1, \gamma_2) \subseteq \Gamma'$ by f). Similarly, $\mathrm{Syz}(\ldots, \hat{\gamma}_i, \ldots) \subseteq \Gamma'$ for all i. Thus $\Gamma' = \Gamma$, and Case I is settled.

j) Case II: Let $\{0,1\}$, $\{2,3\}$ be the connected components of $G_\gamma(\delta)$. Then

$$\mathfrak{V}_\gamma(\delta) = \begin{pmatrix} 1 & 1 & 0 & 0 \\ 1 & 1 & 0 & 0 \\ 0 & 0 & 1 & 1 \\ 0 & 0 & 1 & 1 \end{pmatrix}$$

after a suitable renumbering of the δ_j's. By strict admissibility δ_0 or δ_1 belongs to $\mathrm{Mon}(\gamma_2, \gamma_3)$ and therefore is of type $a\gamma_0\gamma_1/(01) = s(23)$. Since $(023) = (123) = 1$, δ_0 or δ_1 is a multiple of $d_1 := (23)\gamma_0\gamma_1/(01)$. Similarly, δ_2 or δ_3 is a multiple of $d_2 := (01)\gamma_2\gamma_3/(23)$. To settle Case II it suffices to prove $\Gamma((d_1, d_2); \gamma) = \Gamma$.

This identity holds for an arbitrary reduced $\gamma = (\gamma_0, \gamma_1, \gamma_2, \gamma_3)$. In the first step one shows that $d_1 - (01)\gamma_2$ and $d_1 - (01)\gamma_3$ belong to $\mathrm{Mon}(\gamma_0, \gamma_1)$ or $d_2 - (23)\gamma_0$ and $d_2 - (23)\gamma_1$ belong to $\mathrm{Mon}(\gamma_2, \gamma_3)$. Otherwise, say, $d_1 - (01)\gamma_2 \notin \mathrm{Mon}(\gamma_0, \gamma_1)$ and $d_2 - (23)\gamma_0 \notin \mathrm{Mon}(\gamma_2, \gamma_3)$. Then

$$\frac{\gamma_0\gamma_1}{(01)^2} - \frac{\gamma_2}{(23)} = \frac{d_1}{(01)(23)} - \frac{\gamma_2}{(23)} \leq \frac{\gamma_0\gamma_1}{(01)^2} - \frac{\gamma_0}{(01)} - \frac{\gamma_1}{(01)}$$

and hence $\gamma_2/(23) \geq \gamma_0/(01) + \gamma_1/(01) > \gamma_0/(01)$. Similarly $\gamma_0/(01) > \gamma_2/(23)$ is obtained. Contradiction.

Now let us assume $d_1 - (01)\gamma_2$, $d_1 - (01)\gamma_3 \in \mathrm{Mon}(\gamma_0, \gamma_1)$. Then $\Gamma((d_1, d_2); \gamma)$ contains syzygies generating Γ:

$$(\cdot, \cdot, (01), 0), \quad (\cdot, \cdot, 0, (01)), \quad (-\gamma_1/(01), \gamma_0/(01), 0, 0), \quad (0, 0, -\gamma_3, \gamma_2).$$

k) Case III: We may assume

$$\mathfrak{V}_\gamma(\delta) = \begin{pmatrix} 1 & 1 & 1 & 0 \\ 1 & 1 & 1 & 0 \\ 0 & 0 & 0 & 1 \\ 0 & 0 & 0 & 1 \end{pmatrix}.$$

The numbers $(12), (02), (01), \gamma_3$ are pairwise coprime. Because of strict admissibility $(12), (02), (01)$ and hence $(12)\gamma_3, (02)\gamma_3, (01)\gamma_3$ divide δ_2 or δ_3. By a) it suffices to prove that

$$\Gamma((K, (01)\gamma_3, (02)\gamma_3); \gamma) = \Gamma, \qquad K := \mathrm{lcm}(\gamma_0, \gamma_1, \gamma_2).$$

This is true for an arbitrary $\gamma = (\gamma_0, \gamma_1, \gamma_2, \gamma_3)$ with $(012) = 1$. By Sect. 3, Suppl. 3, applied to $m := \gamma_3$, $(01)\gamma_3$ or $K - (01)\gamma_3$ belongs to $\mathrm{Mon}(\gamma_0, \gamma_1)$ and $(02)\gamma_3$ or $K - (02)\gamma_3$ belongs to $\mathrm{Mon}(\gamma_0, \gamma_2)$, which yields syzygies $(\cdot, \cdot, \cdot, (01))$ in $\Gamma((01)\gamma_3; \gamma)$ or $\Gamma(K; \gamma)$ and $(\cdot, \cdot, \cdot, (02))$ in $\Gamma((02)\gamma_3; \gamma)$ or $\Gamma(K; \gamma)$. These two syzygies together with $\mathrm{Syz}(\gamma_0, \gamma_1, \gamma_2) \subseteq \Gamma(K; \gamma)$ generate Γ. This settles Case III and finishes the proof of the theorem.

12 The Main Case for Regular Sequences

In this section we prove among other things that in the main case of elimination the generic elimination ideal $\mathfrak{T}_0(S) \subseteq Q = \mathbb{Z}[U_{j\nu}]_{j,\nu}$ is a principal ideal. We put this result into the set-up of a general elimination theory for regular sequences.

First let us recall some concepts and facts about duality theory for regular sequences and graded complete intersections in general.

Let A be an arbitrary noetherian ring and $A[T] = A[T_0, \ldots, T_n]$ the graded polynomial algebra with indeterminates T_i of positive weights γ_i, $i = 0, \ldots, n$. Let $F_0, \ldots, F_n$ and $G_0, \ldots, G_n$ be regular sequences in $A[T]$ consisting of homogeneous polynomials of positive degrees $\delta_j = \deg F_j > 0$ and $\delta'_j = \deg G_j > 0$, respectively. In addition let us assume that the ideal $\mathfrak{b} = (G_0, \ldots, G_n)$ contains the ideal $\mathfrak{c} = (F_0, \ldots, F_n)$. Thus there are homogeneous polynomials C_{ij} of degree $\delta_j - \delta'_i$ with

$$F_j = \sum_{i=0}^{n} C_{ij} G_i, \quad j = 0, \ldots, n.$$

The homogeneous determinant

$$D := \det(C_{ij})$$

of degree $\tau := \sum_{j=0}^{n} \delta_j - \sum_{i=0}^{n} \delta'_i$ has the following properties:

(1)   *The residue class of D in $C := A[T]/\mathfrak{c}$ is independent of the choice of the coefficients C_{ij}.*

(2) $CD = \mathrm{Ann}_C(\mathfrak{b}/\mathfrak{c})$ *and* $\mathfrak{b}/\mathfrak{c} = \mathrm{Ann}_C(CD)$.

See Wiebe's Satz 2 and its proof in [48],§1. For the reader's convenience we give the outlines of a direct proof in Supplement 6 (following [38], (1.2)).

In case $G_j = T_j$, $j = 0, \ldots, n$, we shall write A_{ij} instead of C_{ij},

$$\Delta = \Delta_F^T = \Delta_{F_0, \ldots, F_n}^{T_0, \ldots, T_n}$$

instead of D and σ instead of τ. Thus,

$$\sigma = \sum_{j=0}^{n} \delta_j - \sum_{i=0}^{n} \gamma_i.$$

For simplicity's sake, the residue class of Δ in C will also be denoted by Δ, if this is not misleading. We will call Δ (as an element of C) the s o c l e d e t e r m i n a n t of $F_0, \ldots, F_n$ (with respect to $T_0, \ldots, T_n$).

Now assume that $G_j = T_j$, $j = 0, \ldots, n$, and that the polynomials $F_0, \ldots, F_n$ define a complete intersection, i. e. that C is finite over A. Then

96

the homogeneous component $C_\sigma = A\Delta$ of C is a free A-module of rank 1, and the graded C-module

$$E := \operatorname{Hom}_A(C, A)$$

is a free C-module generated by the linear form

$$\eta = \eta_F^T = \eta_{F_0,\dots,F_n}^{T_0,\dots,T_n} \in E$$

of degree $-\sigma$, which maps Δ to 1. This linear form η is being used to define for every graded C-module V a natural homogeneous C-isomorphism

$$h = h(V) = h_F^T(V) = h_{F_0,\dots,F_n}^{T_0,\dots,T_n}(V) \; : \; \operatorname{Hom}_C(V, C) \longrightarrow \operatorname{Hom}_A(V, A)$$

of degree $-\sigma$, which maps a C-linear form $\varphi : V \to C$ to the A-linear form $\eta \circ \varphi : V \to A$. Obviously, $h(V)$ is functorial in V. The homomorphism $h(C)$ is an isomorphism by definition; that $h(V)$ is bijective is proved using a free C-presentation of V.

An isomorphism $h(V)$ exists also if C is not finite over A, but V is finite over A, instead. Before we come to its construction, let us note that in the general situation we place ourselves in, for any module V over $B := A[T]/\mathfrak{b}$ there is a natural homogeneous B-isomorphism

$$g = g(V) = g_G^F(V) = g_{G_0,\dots,G_n}^{F_0,\dots,F_n}(V) \; : \; \operatorname{Hom}_B(V, B) \longrightarrow \operatorname{Hom}_C(V, C)$$

of degree τ, given by $\psi \mapsto (v \mapsto \psi(v)D)$. That this homomorphism is an isomorphism follows immediately from property (2) of D mentioned above.

12.1 Lemma *Let A be a noetherian ring. For every regular sequence $F_0, \dots, F_n \in A[T]$ of homogeneous polynomials F_j of degree $\delta_j > 0$ and every graded module V over $C := A[T]/(F_0, \dots, F_n)$ which is finite over A, there is a natural homogeneous isomorphism*

$$h = h(V) = h_F^T(V) \; : \; \operatorname{Hom}_C(V, C) \longrightarrow \operatorname{Hom}_A(V, A)$$

of degree $-\sigma = -(\sum_{j=0}^n \delta_j - \sum_{i=0}^n \gamma_i)$, functorial in V, with the following properties:

(1) *If C is finite over A, $h(V)$ coincides with the isomorphism $\varphi \mapsto \eta \circ \varphi$.*

(2) *If $G_0, \dots, G_n$ is another regular sequence with $(F_0, \dots, F_n)$ contained in $(G_0, \dots, G_n)$, and if V is a graded module over $B := A[T]/(G_0, \dots, G_n)$ which is finite over A, then the diagram*

$$
\begin{array}{ccc}
\operatorname{Hom}_B(V, B) & \xrightarrow{\;h_G^T\;} & \operatorname{Hom}_A(V, A) \\
\Big\downarrow{\scriptstyle g_G^F} & & \Big\downarrow{\scriptstyle \mathrm{id}} \\
\operatorname{Hom}_C(V, C) & \xrightarrow{\;h_F^T\;} & \operatorname{Hom}_A(V, A)
\end{array}
$$

is commutative.

Furthermore, h commutes with flat base changes $A \to A'$.

Proof. Since V is finite over A, there is a regular sequence $F_0', \dots, F_n'$ of homogeneous elements annihilating V such that $C' := A[T]/(F_0', \dots, F_n')$ is finite over A (e.g. take a suitable power of T_j for F_j'). There is a regular sequence $H_0, \dots, H_n$ of homogeneous elements in $A[T]$ such that $(H_0, \dots, H_n) \subseteq (F_0, \dots, F_n) \cap (F_0', \dots, F_n')$ (e.g. take a suitable power of F_j for H_j). Then h_F^T is necessarily the composition $h_{F'}^T \circ (g_{F'}^H)^{-1} \circ g_F^H$, which is compatible with flat base changes.

That this isomorphism is independent of the choices of the two auxiliary sequences $F_0', \dots, F_n'$ and $H_0, \dots, H_n$ is a consequence of the following two simple rules concerning the determinants D and the forms η involved. First, if $(F_0, \dots, F_n) \subseteq (G_0, \dots, G_n) \subseteq (H_0 \dots, H_n)$, then $g_G^F(V) \circ g_H^G(V) = g_H^F(V)$ for every module V over $A[T]/(H_0, \dots, H_n)$. Second, if C and B are finite over A then $h_F^T(V) \circ g_G^F(V) = h_G^T(V)$ for every B-module V, cf. [40], Satz 1.1.   □

Now we look into the elimination theory of algebras

$$C = A[T_0, \dots, T_n]/(F_0, \dots, F_n)$$

over noetherian rings A, where $F_0, \dots, F_n$ is a regular sequence of homogeneous polynomials of positive degrees $\delta_0, \dots, \delta_n$. In Section 10 we introduced the C-ideals $\mathfrak{Q} = \mathfrak{Q}(C)$ and $\mathfrak{T} = \mathfrak{T}(C)$ with

$$0 = \mathfrak{Q} \cap \mathfrak{T}$$

in C. The corresponding decomposition in $A[T]$ we denote by

$$(F_0, \dots, F_n) = \mathfrak{Q}^\star \cap \mathfrak{T}^\star.$$

Recall that $\mathfrak{T} = \operatorname{Ann}_C \mathfrak{Q} = \operatorname{Hom}_C(C/\mathfrak{Q}, C)$. The homogeneous component $\mathfrak{T}_0 = \mathfrak{T}_0^\star \subseteq A$ is by definition the elimination ideal $\mathfrak{T}_0(C)$.

12.2 Lemma *For the ideals $\mathfrak{Q}$ and $\mathfrak{T}$ in $C = A[T]/(F_0, \dots, F_n)$ the following holds:*
(1)   *The associated prime ideals of $\mathfrak{Q}$ in C are the prime ideals $\mathfrak{p} \oplus C_+$, where $\mathfrak{p}$ runs through the set $\operatorname{Ass} A$ of associated prime ideals of A.*
(2)   *The ideal $\mathfrak{Q}$ is the A-torsion of the A-algebra C, and $\mathfrak{Q}$ is annihilated by a non-zero-divisor of A (i.e. $\mathfrak{T}_0$ contains a non-zero-divisor of A). In particular, $\mathfrak{Q}_0 = 0$, i.e. $\mathfrak{Q} \subseteq C_+$.*
(3) $\mathfrak{T} = \operatorname{Ann}_C \mathfrak{Q}$ *and* $\mathfrak{Q} = \operatorname{Ann}_C \mathfrak{T}$.
(4)   *Let $\sigma := \sum_{j=0}^n \delta_j - \sum_{i=0}^n \gamma_i$. For every $m > \sigma$ one has $\mathfrak{T}_m = 0$ and $C_m = \mathfrak{Q}_m$, i.e. $\mathfrak{T}$ is a finite torsionfree A-module contained in $\sum_{m=0}^\sigma C_m$ and $C_{(\sigma+1)} = \sum_{m > \sigma} C_m$ is a torsion module over A contained in $\mathfrak{Q}$.*

Proof. Ad (1): Let $\mathfrak{p} \in \operatorname{Ass} A$ and $\mathfrak{P}^\star := \mathfrak{p} \oplus (T_0, \dots, T_n) \subseteq A[T]$. The localization $R := A[T]_{\mathfrak{P}^\star}$ is of homological codimension $n+1$ because $T_0, \dots, T_n$

is a maximal regular sequence in R. On the other hand, $F_0, \ldots, F_n$ is a regular sequence in R, too. Hence $\mathfrak{P}^*$ is associated to $(F_0, \ldots, F_n)$, i. e. $\mathfrak{P} = \mathfrak{p} \oplus C_+$ is associated to $\mathfrak{Q}$.

Conversely, let $\mathfrak{P}^* = \mathfrak{p} \oplus (T_0, \ldots, T_n)$ be an associated prime ideal of $\mathfrak{Q}^*$ and thus of $(F_0, \ldots, F_n)$. Then $F_0, \ldots, F_n$ is a maximal regular sequence in $R := A[T]_{\mathfrak{P}^*}$. Since $T_0, \ldots, T_n$ is a regular sequence in R of same length, $\mathfrak{P}^*$ is associated to $(T_0, \ldots, T_n)$, which means that $\mathfrak{p} \in \text{Ass}\, A$.

Ad (2): Let K denote the total ring of fractions of A. By 7.3 and 7.4, $K \otimes_A C$ is finite over K, i. e. all associated prime ideals of $K \otimes_A C$ contain $K \otimes_A C_+$. But these are exactly the extensions of the prime ideals belonging to $\mathfrak{Q}$ by part (1). Therefore $K \otimes_A \mathfrak{Q} = 0$, i. e. $\mathfrak{Q}$ is part of the A-torsion of C. The reversed inclusion follows directly from (1).

Ad (3): The first equality is true in general. The non-trivial inclusion $\mathfrak{Q} \supseteq \text{Ann}_C \mathfrak{T}$ of the second one follows from (2).

Ad (4): Let K denote the total ring of fractions of A, and consider $m > \sigma$. By 7.4, $K \otimes_A C_m = (K \otimes_A C)_m = 0$, which yields $C_m \subseteq \mathfrak{Q}$ by (2). Finally, $\mathfrak{T}_m = \mathfrak{T} \cap C_m = \mathfrak{T} \cap \mathfrak{Q}_m = 0$. $\qquad\square$

Using the results from duality theory we can now prove the following theorem, which plays a central role in our considerations.

12.3 Theorem *Let A be a noetherian ring, $F_0, \ldots, F_n$ be a regular sequence in $A[T]$ of homogeneous polynomials of positive degrees $\delta_0, \ldots, \delta_n$, $C = A[T]/(F_0, \ldots, F_n)$, $\sigma = \sum_{j=0}^n \delta_j - \sum_{i=0}^n \gamma_i$ and $\Delta = \Delta_F^T \in C_\sigma$. For the ideals $\mathfrak{Q}$ and $\mathfrak{T}$ the following holds:*

(1) *In degree σ, $\mathfrak{T}_\sigma = A\Delta \cong A$.*

(2) *There is a canonical isomorphism*

$$\mathfrak{T} = \text{Hom}_C(C/\mathfrak{Q}, C) \longrightarrow \text{Hom}_A(C/\mathfrak{Q}, A)(-\sigma)$$

of graded C-modules. In particular, for every $m \in \mathbb{N}$ there is a canonical A-isomorphism

$$\mathfrak{T}_m \longrightarrow \text{Hom}_A(C_{\sigma-m}/\mathfrak{Q}_{\sigma-m}, A) = \text{Hom}_A(C_{\sigma-m}, A) = C_{\sigma-m}^*.$$

(3) *In every degree m, the isomorphism $\mathfrak{T}_m \to C_{\sigma-m}^*$ from (2) runs as follows:*

$$t_m \mapsto \left(c_{\sigma-m} \mapsto \frac{t_m c_{\sigma-m}}{\Delta} \right)$$

where $t_m \in \mathfrak{T}_m$, $c_{\sigma-m} \in C_{\sigma-m}$.

(4) *In degree 0, the inverse of the isomorphism $\mathfrak{T}_0 \to C_\sigma^*$ is given by $\psi \mapsto \psi(\Delta)$, and*

$$\mathfrak{T}_0 = \text{Ann}_A(C_\sigma/A\Delta).$$

Proof. To prove (1), let us note first that $(T_0, \ldots, T_n) \cdot \mathfrak{T}_\sigma \subseteq C_{(\sigma+1)} \cap \mathfrak{T} \subseteq \mathfrak{Q} \cap \mathfrak{T} = 0$ by 12.2(4). Thus, $\mathfrak{T}_\sigma \subseteq \text{Ann}_C(T_0, \ldots, T_n) = C\Delta$, i. e. $\mathfrak{T}_\sigma \subseteq A\Delta$.

Conversely, we have $\Delta \cdot \mathfrak{Q} \subseteq \Delta \cdot (T_0, \ldots, T_n) = 0$ by 12.2(2) and therefore $\Delta \in \text{Ann}_C \mathfrak{Q} = \mathfrak{T}$, i.e. $\Delta \in \mathfrak{T}_\sigma$. This shows $\mathfrak{T}_\sigma = A\Delta$. Finally, $A\Delta \cong A$ because of $A \cap \text{Ann}_C \Delta = A \cap (T_0, \ldots, T_n) = 0$.

Part (2) follows from Lemma 12.1 applied to the C-module $V := C/\mathfrak{Q}$, which is a finite A-module. That the A-duals of $C_{\sigma-m}/\mathfrak{Q}_{\sigma-m}$ and $C_{\sigma-m}$ coincide, follows from the fact that $\mathfrak{Q}_{\sigma-m}$ is a torsion module by 12.2(2).

Ad (3): Fix $m \in \mathbb{N}$. Note that $t_m c_{\sigma-m} \in \mathfrak{T}_\sigma$. By (1), therefore, there is a unique $a \in A$ such that $t_m c_{\sigma-m} = a\Delta$, and the homomorphism described in (3) proves to be well-defined. To check that it equals the canonical isomorphism we may, because $C^*_{\sigma-m}$ is torsionfree, go over to the total ring of fractions of A, that is, we may assume that C is finite over A. Then let $\eta = \eta^T_F$ denote the canonical generator of the C-module C^*. By Lemma 12.1(1), the canonical isomorphism maps t_m to $\eta \circ (c_{\sigma-m} \mapsto t_m c_{\sigma-m})$, i.e. to the mapping $c_{\sigma-m} \mapsto \eta(t_m c_{\sigma-m})$. But $\eta(t_m c_{\sigma-m}) = t_m c_{\sigma-m}/\Delta$ because of $\eta(\Delta) = 1$.

Ad (4): $t_0 \in \mathfrak{T}_0$ corresponds to $\psi = (c_\sigma \mapsto t_0 c_\sigma/\Delta) \in C^*_\sigma$. Hence $\psi(\Delta) = t_0$. The inclusion $\mathfrak{T}_0 \subseteq \text{Ann}_A(C_\sigma/A\Delta)$ follows from $\mathfrak{T}_0 C_\sigma \subseteq \mathfrak{T}_\sigma$ and $\mathfrak{T}_\sigma = A\Delta$ by (1). Conversely, let $a \in A$ be an element with $aC_\sigma \subseteq A\Delta$. Then $\psi : c_\sigma \mapsto ac_\sigma/\Delta$ is an A-linear form $C_\sigma \to A$ with $\psi(\Delta) = a$. Therefore $a \in \mathfrak{T}_0$. □

12.4 Corollary *Let $F_0, \ldots, F_n$ be a regular sequence in $A[T]$ as in 12.3 and $C = A[T]/(F_0, \ldots, F_n)$. If A is an integrally closed domain, the elimination ideal $\mathfrak{T}_0$ of C is a divisorial ideal. In particular, if A is factorial, $\mathfrak{T}_0$ is a principal ideal $\neq 0$.*

Proof. The result follows simply from the fact that by 12.3(2) the ideal $\mathfrak{T}_0$ is the dual of $C_\sigma/\mathfrak{Q}_\sigma \neq 0$. □

Note that in the proof of 12.4 just the existence of an isomorphism $\mathfrak{T}_0 \cong C^*_\sigma$ is being used and not an explicit description as in 12.3(3). Thus the proof of 12.4 depends merely on the first section of the proof of 12.1.

12.5 Corollary *Let $F_0, \ldots, F_n$ be a regular sequence in $A[T]$ as in 12.3 and $C = A[T]/(F_0, \ldots, F_n)$. The following conditions are equivalent:*

(1) C *is finite over A, i.e. C is a complete intersection over A.*
(2) $\mathfrak{T}_0(C) = A$.
(3) $C_\sigma = A\Delta$.
(3′) $C_\sigma = A\Delta + \mathfrak{Q}_\sigma$.

Proof. Only the implication (3′) ⇒ (2) is not obvious. Because $A\Delta \cong A$ and $\mathfrak{Q}_\sigma$ is a torsion module, (3′) implies $C_\sigma = A\Delta \oplus \mathfrak{Q}_\sigma$. Therefore there is a linear form $\psi \in C^*_\sigma$ with $\psi(\Delta) = 1$. This implies $1 \in \mathfrak{T}_0$ by 12.3(4). □

12.5 can be partially generalized as follows.

12.6 Theorem *Assume that A is a noetherian ring, which is not the zero ring. Let $F_0, \ldots, F_n$ be arbitrary homogeneous polynomials in $A[T]$ of positive degrees $\delta_0, \ldots, \delta_n$ and $\Delta \in C = A[T]/(F_0, \ldots, F_n)$ the (residue class of the) determinant obtained from representations $F_j = \sum_{i=0}^{n} A_{ij} T_i$ in $A[T]$. Equivalent are:*

(1) *C is finite over A, i.e. C is a complete intersection over A.*
(2) *$\mathfrak{T}_0(C) = A$.*
(3) *$C_\sigma = A\Delta$ and $C_\sigma \neq 0$.*

Proof. The equivalence of conditions (1) and (2) follows from 11.2 and 10.2. Therefore it remains to prove (3) $\Rightarrow$ (2). Let $A \neq 0$. We consider C as a specialization $Q \to A$ of the Q-algebra $S = Q[T]/(F_0, \ldots, F_n)$ where the F_j are the generic homogeneous polynomials

$$F_j = \sum_{\nu_0 \gamma_0 + \cdots + \nu_n \gamma_n = \delta_j} U_{j\nu} T^\nu$$

over the polynomial ring $Q = \mathbb{Z}[U_{j\nu}]_{j,\nu}$. Assume at first that $F_0, \ldots, F_n$ is a regular sequence in $Q[T]$. Then S_σ is a Q-module of positive rank ($= 1$). Therefore, for every specialization $Q \to A'(\neq 0)$ the A'-module C'_σ in $C' := A' \otimes_Q S$ is not zero. (Compare $\text{Ann}_Q S_\sigma$ to $\text{Ann}_{A'} C'_\sigma$.) So condition (3) carries over to every specialization $A' = A/\mathfrak{m}$, $\mathfrak{m}$ maximal in A. (2) may be checked over any non-zero specialization. Thus we may assume that A is a field k. To prove (3) $\Rightarrow$ (2) in the special situation $A = k$, we have to show that $C_{(T_0, \ldots, T_n)}$ is a complete intersection of dimension 0. This however follows from Satz 1 in [48] (cf. Theorem 2.3.16 in [6]) since $\Delta \neq 0$ belongs to the Fitting ideal of the maximal ideal of $C_{(T_0, \ldots, T_n)}$.

Finally, to show that $F_0, \ldots, F_n$ indeed form a regular sequence in $Q[T]$ we verify condition (3) of Lemma 8.1. Let $L = A/\mathfrak{m}$, where $\mathfrak{m}$ is a maximal ideal in A such that $C_\sigma/\mathfrak{m}C_\sigma \neq 0$. Again by Wiebe's Satz we get that $L \otimes_Q S = L \otimes_A C$ is finite over L. $\qquad\square$

Supplements

1. a) In the situation of Theorem 12.3 the following assertions hold for every $m \in \mathbb{Z}$:
(1) $\mathfrak{Q}_m = \{c_m \in C_m : c_m \mathfrak{T}_{\sigma-m} = 0\}$.
(2) $\mathfrak{Q}_m = \{c_m \in C_m : c_m C_{\sigma-m} \subseteq \mathfrak{Q}_\sigma\}$.
(3) $\mathfrak{Q}_m + \mathfrak{T}_m = \{c_m \in C_m : c_m C_{\sigma-m} \subseteq \mathfrak{Q}_\sigma + \mathfrak{T}_\sigma\}$.
(For the proof of (1) and (2) consider $Q(A) \otimes_A C$ which is a complete intersection over the total quotient ring $Q(A)$ of A. For (3) let $c_m \in C_m$ be an element with $c_m C_{\sigma-m} \subseteq \mathfrak{Q}_\sigma + \mathfrak{T}_\sigma$ and $a \in A$ any non-zero-divisor which annihilates $\mathfrak{Q}_\sigma$. Then

$ac_m C_{\sigma-m} \subseteq a\mathfrak{T}_\sigma = a\Delta A$ and the isomorphism $\mathfrak{T}_m \to C^*_{\sigma-m}$ provides an element $t_m \in \mathfrak{T}_m$ with $(ac_m - at_m)C_{\sigma-m} = 0$.)

b) As a corollary of (3) in a) one obtains: If A is a Dedekind domain or, more generally, a Gorenstein ring of dimension 1, then $C/(\mathfrak{Q}+\mathfrak{T})$ is a (graded) semilocal zero-dimensional Gorenstein ring. ($C_\sigma/(\mathfrak{Q}_\sigma + \mathfrak{T}_\sigma) = (C_\sigma/\mathfrak{Q}_\sigma)/A\Delta$, and $C_\sigma/\mathfrak{Q}_\sigma$ is a torsionless A-module of rank 1 with the free submodule $A\Delta$, therefore the A-socle of $C_\sigma/(\mathfrak{Q}_\sigma + \mathfrak{T}_\sigma)$ is one-dimensional if A is local.)

2.† The converse of the isomorphism

$$\mathfrak{T} \longrightarrow \mathrm{Hom}_A(C/\mathfrak{Q}, A)(-\sigma) = \mathrm{Hom}_A(C, A)(-\sigma) = (C^*)(-\sigma)$$

of graded C-modules in Theorem 12.3 can be given in the following concrete way which generalizes 12.3(4), where this isomorphism is treated in degree 0.

Consider a representation

$$F_j \otimes 1 - 1 \otimes F_j = \sum_{i=0}^n C_{ij}(T_i \otimes 1 - 1 \otimes T_i)$$

in $A[T]^e := A[T] \otimes_A A[T]$ with homogeneous elements C_{ij} of degree $\delta_j - \gamma_i$. Then $D^e := \det(C_{ij})$ is homogeneous of degree σ. By

$$\Delta^e$$

we denote the residue class of D^e in $C^e := C \otimes_A C$. This residue class is uniquely determined since the $F_j \otimes 1 - 1 \otimes F_j$ form a regular sequence by Supplement 6a) of Section 7. Therefore even the residue class of D^e modulo this sequence is unique. Furthermore, $\Delta^e \in C^e$ is symmetric, i.e. invariant under the automorphism $b \otimes c \mapsto c \otimes b$ of C^e.

Δ^e annihilates the kernel $\mathfrak{I}$ of the multiplication mapping $C^e \to C$ which is generated by the residue classes of the $T_i \otimes 1 - 1 \otimes T_i$.

Let κ denote the natural A-homomorphism

$$\kappa : C^e \longrightarrow \mathrm{Hom}_A(C^*, C)$$

with $c \otimes d \mapsto (\psi \mapsto \psi(c)d)$ which is homogeneous of degree 0 (and an isomorphism if C is finite over A). For every element w in the annihilator of $\mathfrak{I}$ the homomorphism $\kappa(w)$ is even C-linear. Then

$$\Theta := \kappa(\Delta^e) \; : \; (C^*)(-\sigma) \longrightarrow C$$

induces the converse $(C^*)(-\sigma) \to \mathfrak{T}$ of the canonical isomorphism $\mathfrak{T} \to (C^*)(-\sigma)$. (First one shows that Θ maps C^* into $\mathfrak{T} = \mathrm{Ann}_C\mathfrak{Q}$. Let $\psi \in C^*$ and $q \in \mathfrak{Q}$ be given. It suffices to show that $q\Theta(\psi) = 0$. But $q\Theta(\psi) = \Theta(q\psi) = 0$ because ψ vanishes on $qC \subseteq \mathfrak{Q}$ such that $q\psi = 0$.

Now the assertion can be checked over the total ring of quotients of A because $\mathfrak{T}$ is torsionfree over A. Thus we may assume that A coincides with its total ring of quotients. Then C is a finite complete intersection over A. But then the result is well known from duality theory for finite complete intersections. In this case Θ maps the standard C-base element $\eta \in (C^*)_\sigma$ of C^* (with $\eta(\Delta) = 1$) to $1 \in C$, cf. Sect. 3 of [38].)

Note that the result yields a method to compute the ideal $\mathfrak{T}$ of inertia forms.

3. Let A denote the Dedekind domain $\mathbb{Z}[\sqrt{-5}]$. Then $F_0 := 2T_0^3 + (1+\sqrt{-5})T_1^2$, $F_1 := T_0^6 + (1 - \sqrt{-5})T_0^3 T_1^2 + 3T_1^4$ is a regular sequence in $A[T] = A[T_0, T_1]$. The homogeneous parts of $\mathfrak{T}$ are not free over A in general, neither are the homogeneous parts of $C/\mathfrak{Q}$, $C = A[T]/(F_0, F_1)$. (Consider C_6, C_7 and $\mathfrak{T}_6$, $\mathfrak{T}_7$, for instance. Cf. Sect. 13, Suppl. 1c).)

4.† a) The canonical duality $\mathfrak{T}_m \times C_{\sigma-m} \to A$ in Theorem 12.3 given by $(t_m, c_{\sigma-m}) \mapsto t_m c_{\sigma-m}/\Delta$ yields the exact sequence

$$0 \longrightarrow \mathfrak{Q}_{\sigma-m} \longrightarrow C_{\sigma-m} \xrightarrow{\alpha_m} \mathfrak{T}_m^*,$$

where α_m is defined by

$$c_{\sigma-m} \mapsto \left(t_m \mapsto \tfrac{t_m c_{\sigma-m}}{\Delta}\right).$$

The bidual of α_m is the dual of the isomorphism $\mathfrak{T}_m \to C_m^*$ in 12.3(3). In particular, α_m is surjective if A is a Dedekind domain or, more generally, a Gorenstein ring of dimension 1.

b) In general, α_m is not surjective. (Almost any concrete situation will provide examples. For instance, for the generic polynomials

$$XT_0^3 + YT_1^2, \quad UT_0^6 + VT_0^3T_1^2 + WT_1^4$$

over $A := Q := \mathbb{Z}[X, Y, U, V, W]$ with $\deg T_0 := 2$, $\deg T_1 := 3$ the Q-module $C_6 \cong Q^2/Q(X, Y)$ is torsionfree of rank 1 but not free and therefore not reflexive, whence $\alpha_7 : C_6 \to \mathfrak{T}_7^* \cong Q$ is not surjective.)

5.† Let A be a Dedekind domain or, more generally, a Gorenstein ring of dimension 1 and $F_0, \ldots, F_n$ a regular sequence in $A[T]$ of homogeneous polynomials; adopt the usual conventions about $\sigma, C, \mathfrak{Q}, \mathfrak{T}$. Then for every $m \in \mathbb{Z}$ the following holds:

$$\operatorname{length}_A C_m/(\mathfrak{Q}_m + \mathfrak{T}_m) = \operatorname{length}_A C_{\sigma-m}/(\mathfrak{Q}_{\sigma-m} + \mathfrak{T}_{\sigma-m}).$$

(For every $m \in \mathbb{Z}$, let ι_m denote the inclusion $\mathfrak{T}_m \hookrightarrow C_m$. The diagram

$$
\begin{array}{ccc}
\mathfrak{T}_m & \longrightarrow & C_{\sigma-m}^* \\
\downarrow{\iota_m} & & \downarrow{\iota_{\sigma-m}^*} \\
C_m & \xrightarrow{\alpha_{\sigma-m}} & \mathfrak{T}_{\sigma-m}^*
\end{array}
$$

with canonical homomorphisms is commutative. Here $\mathfrak{T}_m \to C_{\sigma-m}^*$ is the canonical isomorphism of 12.3(3) and $\alpha_{\sigma-m}$ the surjection treated in Suppl. 4. As a consequence $C_m/(\mathfrak{Q}_m + \mathfrak{T}_m)$ is isomorphic to the cokernel of $\iota_{\sigma-m}^*$, which is seen to have the length of $C_{\sigma-m}/(\mathfrak{Q}_{\sigma-m} + \mathfrak{T}_{\sigma-m})$.)

6. Let $f_0, \ldots, f_n$ and $g_0, \ldots, g_n$ be regular sequences in the noetherian ring R such that the ideal $\mathfrak{b} := (g_0, \ldots, g_n)$ contains the ideal $\mathfrak{c} := (f_0, \ldots, f_n)$. Furthermore, choose elements c_{ij} in R such that $f_j = \sum_{i=0}^n c_{ij}g_i$, $j = 0, \ldots, n$. Let $d := \det(c_{ij})$ in R and denote $R/\mathfrak{c}$ by R', $\mathfrak{b}/\mathfrak{c}$ by $\mathfrak{b}'$. Then:

(1)   The residue class d' of d in R' is independent of the choice of the coefficients c_{ij} and generates the Fitting ideal $\mathfrak{f}_{A'}(\mathfrak{b}') = \mathfrak{f}_{0,A'}(\mathfrak{b}')$.

(2) $R'd' = \operatorname{Ann}_{A'}(\mathfrak{b}')$ and $\mathfrak{b}' = \operatorname{Ann}_{A'}(R'd')$.

For the proof we may assume that R is local and that $f_0, \ldots, f_n$ and $g_0, \ldots, g_n$ are (strongly) regular sequences contained in the maximal ideal $\mathfrak{m}$ of R.

To prove (1) take another representation $f_j = \sum_{i=0}^n \tilde{c}_{ij}g_i$, $j = 0, \ldots, n$, and $\tilde{d} := \det(\tilde{c}_{ij})$. Proceeding step by step we may assume $\tilde{c}_{ij} = c_{ij}$ for $j = 1, \ldots, n$. Then $d - \tilde{d} = \det \mathfrak{M}$, where $\mathfrak{M} = (d_{ij})$ with $d_{ij} := c_{ij}$ for $j > 0$ and $d_{i0} := c_{i0} - \tilde{c}_{i0}$. By Cramer's rule $g_i \det \mathfrak{M} \subseteq (f_1, \ldots, f_n)$ for all i, thus $f_0 \det \mathfrak{M} = f_0(d - \tilde{d})$ belongs to $(f_1, \ldots, f_n)$ which implies $d - \tilde{d} \in (f_1, \ldots, f_n) \subseteq \mathfrak{c}$.

To prove the second part of (1) we compute the Fitting ideal $\mathfrak{f}_A(\mathfrak{b}')$ of $\mathfrak{b}'$ as an R-module. $\mathfrak{b}'$ has the representation $\mathfrak{b}' = R^{n+1}/U$, where U is generated by the regular syzygies $g_k e_i - g_i e_k$, $i < k$, and the elements $\sum_{i=0}^n c_{ij}e_i$, $j = 0, \ldots, n$.

Therefore $\mathfrak{f}_A(\mathfrak{b}')$ is generated by d and elements belonging to $\mathfrak{c}$ (using an argument as above in the first part). $\mathfrak{f}_{A'}(\mathfrak{b}')$ is the homomorphic image of $\mathfrak{f}_A(\mathfrak{b}')$ in R'.

(2) is proved by induction on n. For this we change the generators $g_0, \ldots, g_n$ of $\mathfrak{b}$ (using (1)) and assume that g_0 is prime to $\mathfrak{e} := (f_1, \ldots, f_n)$. (By an old trick of E.D. Davis it is enough to replace g_0 by $h_m := g_0 + f_0^m$ for a suitable $m \geq 2$, since for every $\mathfrak{p} \in \mathrm{Ass}\,(R/\mathfrak{e})$ at most one of the elements h_m, $m \in \mathbb{N}^*$, belongs to $\mathfrak{p}$ because of $h_m - h_{m+r} = f_0^m(1 - f_0^r)$ for $m, r \in \mathbb{N}^*$.)

Let $K := R/\mathfrak{e}$. By induction hypothesis applied to the regular sequences $f_1, \ldots, f_n$ and $g_1, \ldots, g_n$ in R/Rg_0 we have $(K/Kg_0)d_0 = \mathrm{Ann}_{K/Kg_0}(K\mathfrak{b}/Kg_0) = (Kg_0 : K\mathfrak{b})/Kg_0$ and $K\mathfrak{b}/Kg_0 = \mathrm{Ann}_{K/Kg_0}((K/Kg_0)d_0) = (Kg_0 : Kd_0)/Kg_0$ where $d_0 := \det(c_{ij})_{1 \leq i, j \leq n}$. From the last equality we get $K\mathfrak{b} = Kg_0 : Kd_0$. By Cramer's rule $d_0 f_0 \equiv g_0 d \bmod \mathfrak{e}$, hence $Kf_0 : Kd = Kg_0 : Kd_0 = K\mathfrak{b}$ and

$$K\mathfrak{b}/Kf_0 \;=\; (Kf_0 : Kd)/Kf_0 \;=\; \mathrm{Ann}_{K/Kf_0}((K/Kf_0)d)\,.$$

The canonical isomorphism from $(Kg_0 : K\mathfrak{b})/Kg_0$ onto $(Kf_0 : K\mathfrak{b})/Kf_0$ maps the residue class of d_0 to the residue class of d. Therefore

$$(K/Kf_0)d_0 \;=\; (Kf_0 : K\mathfrak{b})/Kf_0 \;=\; \mathrm{Ann}_{K/Kf_0}(K\mathfrak{b}/Kf_0)\,.$$

But $K/Kf_0 = R'$, $K\mathfrak{b}/Kf_0 = \mathfrak{b}'$ and $(K/Kf_0)d = R'd'$.

7.[†] Let A be a noetherian ring of dimension ≤ 1 and $F_0, \ldots, F_n$ a sequence of homogeneous polynomials of positive degrees in $A[T] = A[T_0, \ldots, T_n]$; adopt the usual conventions about weights, C, $\mathfrak{X}$. The sequence $F_0, \ldots, F_n$ is regular in $A[T]$ if and only if the elimination ideal $\mathfrak{X}_0$ contains a non-zero-divisor of A and all fibres of $\mathrm{Proj}\,C \to \mathrm{Spec}\,A$ are finite.

(Let $F_0, \ldots, F_n$ be regular. By 12.2(2) $\mathfrak{X}_0$ contains a non-zero-divisor. Finiteness of all fibres means that for all maximal ideals $\mathfrak{m} \subseteq A$ containing $\mathfrak{X}_0$ the dimension of $C/\mathfrak{m}C$ is ≤ 1, which is clear because of $\dim C \leq 1$.

Conversely, assume that the finiteness condition for fibres holds and that $\mathfrak{X}_0$ contains a non-zero-divisor $f \in A$. Furthermore, we may assume that A is local. If f is a unit in A then C is even a finite complete intersection by 11.2. Thus, from now on, let f be a non-zero-divisor in the maximal ideal $\mathfrak{m}$ of A. Then $A[T]$ is a Cohen-Macaulay ring of dimension $n + 2$. Let $\mathfrak{P}$ be any homogeneous prime ideal of $A[T]$ containing $F_0, \ldots, F_n$. It is enough to show that $\mathrm{codim}\,\mathfrak{P} \geq n + 1$. If $f \notin \mathfrak{P}$ then $A[T]_{\mathfrak{P}}$ is a localization of $A_f[T]$, where the sequence defines a complete intersection by 11.2. Thus $\mathrm{codim}\,\mathfrak{P} \geq n+1$. Now let $f \in \mathfrak{P}$ and therefore $\mathfrak{m} \subseteq \mathfrak{P}$. Assume $\mathrm{codim}\,\mathfrak{P} \leq n$. Then $\dim A[T]/\mathfrak{P} \geq 2$ which is impossible because $\mathfrak{P}$ belongs to the fibre of $\mathrm{Proj}\,C \to \mathrm{Spec}\,A$ over $\{\mathfrak{m}\}$.)

IV Resultants

13 Resultant Ideals

We place ourselves in the situation discussed in the last section, now following [43] closely. We consider a regular sequence $F_0, \ldots, F_n$ of homogeneous
polynomials in $A[T] = A[T_0, \ldots, T_n]$ of positive degrees $\delta_0, \ldots, \delta_n$ over a
noetherian ring A and the A-algebra $C = A[T]/(F_0, \ldots, F_n)$. Using the homogeneous ideal $\mathfrak{T}(C)$ of inertia forms, we define the graded residue class
algebra
$$\bar{C} := C/\mathfrak{T}(C)$$
which has the same projective spectrum as C. The weights of the inde-
terminates $T_0, \ldots, T_n$ are $\gamma_0, \ldots, \gamma_n$. Then the socle determinant $\Delta = \Delta_F^T$
belongs to C_σ, $\sigma = \sum_{j=0}^n \delta_j - \sum_{i=0}^n \gamma_i$, cf. Section 12.

Theorem 12.3 and its corollaries show the importance of the homogeneous part
$$\bar{C}_\sigma = C_\sigma/\mathfrak{T}_\sigma = C_\sigma/A\Delta = A[T]_\sigma/((F_0, \ldots, F_n)_\sigma + A\Delta)$$
of degree σ in $\bar{C}$. By 12.3(4), $\mathfrak{T}_0$ is the annihilator of $\bar{C}$. However, annihilator ideals lack good functorial properties. *For this reason we favour
instead of $\mathfrak{T}_0 = \mathrm{Ann}_A \bar{C}_\sigma$ the Fitting ideal $\mathfrak{f}_0(\bar{C}_\sigma)$ which has the same zero
set as $\mathfrak{T}_0$ in* $\mathrm{Spec}\, A$.

If A is an integrally closed domain the ideal $\mathfrak{T}_0 \cong C_\sigma^*$ is a divisorial
ideal. Thus even the divisorial part of $\mathfrak{f}_0(\bar{C}_\sigma)$ has the same zero set as $\mathfrak{T}_0$.
To describe this divisor we start with the following lemma:

13.1 Lemma *Let A be an integrally closed noetherian domain. Then:*
*(1) The A-submodules $\mathfrak{T}_\sigma = A\Delta$ and $\mathfrak{T}_0 C_\sigma$ of C_σ coincide in codimen-
sion 1.*
(2) The canonical homomorphism
$$(C_\sigma/\mathfrak{Q}_\sigma)/\mathfrak{T}_0(C_\sigma/\mathfrak{Q}_\sigma) \longrightarrow (C_\sigma/\mathfrak{Q}_\sigma)/\mathfrak{T}_\sigma = \bar{C}_\sigma/\mathfrak{Q}_\sigma$$
is an isomorphism in codimension 1.

Proof. We may assume that A is a discrete valuation ring. Then C_σ splits
into its torsion submodule $\mathfrak{Q}_\sigma$ and a free submodule Ac of rank 1. Let ψ be
the linear form on C_σ with $\psi(c) = 1$. Then $C_\sigma^* = A\psi$ and $\mathfrak{T}_0 = A\psi(\Delta) = Aa$
by 12.3(4), where $\Delta = u + ac$, $u \in \mathfrak{Q}_\sigma$. Because of $\mathfrak{T}_0 C_\sigma \subseteq \mathfrak{T}_\sigma = A\Delta$ there

is an element $b \in A$ with $ac = b(u + ac)$. From this follows $b = 1$, $u = 0$ and thus $\mathfrak{T}_0 C_\sigma = aC_\sigma = A\Delta$.   □

In general, for an arbitrary ideal $\mathfrak{a} \neq 0$ in an integrally closed noetherian domain A we denote by

$$\mathfrak{div}\,\mathfrak{a}$$

the corresponding divisorial ideal, which is the double dual of $\mathfrak{a}$, and by $\operatorname{div}\mathfrak{a}$ its divisor. For an arbitrary finite torsion module M over A

$$\chi(M) := \sum_{\operatorname{ht}\mathfrak{p}=1} \operatorname{length}_{A_\mathfrak{p}}(M_\mathfrak{p}) \cdot \mathfrak{p} = \operatorname{div}\mathfrak{f}_0(M)$$

is the divisor associated to M in the sense of Bourbaki [4], Ch. VII, § 4, no. 7. The divisorial ideal $\mathfrak{div}\,\mathfrak{f}_0(M)$ belonging to the divisor $\chi(M)$ will be denoted simply by $\mathfrak{d}(M)$. We have $\operatorname{div}\mathfrak{a} = \chi(A/\mathfrak{a}) = \chi(A/\mathfrak{div}\,\mathfrak{a})$ and $\mathfrak{div}\,\mathfrak{a} = \mathfrak{d}(A/\mathfrak{a}) = \mathfrak{d}(A/\mathfrak{div}\,\mathfrak{a})$.

The divisor $\operatorname{div}\mathfrak{f}_0(\bar{C}_\sigma)$ has the same support as the divisor $\operatorname{div}\mathfrak{T}_0$ of the elimination ideal $\mathfrak{T}_0 = \operatorname{Ann}_A\bar{C}_\sigma$. Furthermore we have

$$\operatorname{div}\mathfrak{T}_0 \leq \chi(\bar{C}_\sigma)$$

(Cayley-Hamilton), more precisely:

13.2 Lemma *Let A be an integrally closed noetherian domain. Then*

$$\chi(\bar{C}_\sigma) = \chi(\mathfrak{Q}_\sigma) + \chi(A/\mathfrak{T}_0) = \chi(\mathfrak{Q}_\sigma) + \operatorname{div}\mathfrak{T}_0 = \operatorname{div}\mathfrak{f}_1(C_\sigma) + \operatorname{div}\mathfrak{T}_0 .$$

Proof. $C_\sigma/\mathfrak{Q}_\sigma$ is a finite torsionfree A-module of rank one. Therefore

$$\chi((C_\sigma/\mathfrak{Q}_\sigma)/\mathfrak{T}_0 \cdot (C_\sigma/\mathfrak{Q}_\sigma)) = \chi(A/\mathfrak{T}_0).$$

The first equality follows now from 13.1(2). By definition $\chi(A/\mathfrak{T}_0) = \operatorname{div}\mathfrak{T}_0$. It remains to prove the equality $\chi(\mathfrak{Q}_\sigma) = \operatorname{div}\mathfrak{f}_1(C_\sigma)$. But $\chi(\mathfrak{Q}_\sigma) = \operatorname{div}\mathfrak{f}_0(\mathfrak{Q}_\sigma) = \operatorname{div}\mathfrak{f}_1(C_\sigma)$ because $C_\sigma/\mathfrak{Q}_\sigma$ is torsionfree of rank one.   □

Further information about the divisor $\chi(\bar{C}_\sigma)$ can be derived from the explicit finite free A-resolution

$$0 \longrightarrow V_{n+1} \longrightarrow V_n \longrightarrow \cdots \longrightarrow V_1 \longrightarrow V_0 \overset{\varepsilon}{\longrightarrow} C_\sigma \longrightarrow 0$$

of C_σ which is the homogeneous part of degree σ of the Koszul resolution of $C = A[T]/(F_0,\ldots,F_n)$. Because of $\bar{C}_\sigma = C_\sigma/A\Delta$ and $A\Delta \cong A$, $\bar{C}_\sigma$ has a similar finite free resolution. The mere existence of such a resolution implies the following result, cf. [4], Ch. VII, § 4, no. 7, Cor. 2 to Prop. 16:

13.3 Theorem *The divisor $\chi(\bar{C}_\sigma)$ is a principal divisor.*

It should be noted that the divisorial ideal $\mathfrak{T}_0$ need not be principal, cf. Supplement 1 for examples.

We call the principal ideal of the integrally closed noetherian domain A belonging to the principal divisor $\chi(\bar{C}_\sigma)$ the r e s u l t a n t i d e a l of the polynomials $F_0, \ldots, F_n \in A[T]$ (or of the algebra $C = A[T]/(F_0, \ldots, F_n)$ represented by $F_0, \ldots, F_n$) and denote it by

$$\mathfrak{R} = \mathfrak{R}(F_0, \ldots, F_n) = \mathfrak{R}_A(F_0, \ldots, F_n).$$

$\mathfrak{R}$ is the divisorial ideal $\mathfrak{d}(\bar{C}_\sigma) = \mathfrak{div}\, \mathfrak{f}_0(\bar{C}_\sigma) \subseteq A$.

In case A is not only integrally closed in its quotient field, but even a factorial domain, a generator of $\mathfrak{R}$ can be constructed as the greatest common divisor of the elements in an arbitrary set of generators of the (determinantal) ideal $\mathfrak{f}_0(\bar{C}_\sigma)$.

In Section 15 we will construct a canonical generator $R = \mathrm{R}(F_0, \ldots, F_n)$ of $\mathfrak{R}$ which will be the resultant of $F_0, \ldots, F_n$. Here and in the next section we want to study more closely the relations between the resultant ideal $\mathfrak{R}$ and the elimination ideal $\mathfrak{T}_0$.

$\mathfrak{R}$ is contained in $\mathfrak{T}_0$, their divisors have the same support and therefore both ideals have the same set of zeros. By 13.2, $\mathfrak{R} = \mathfrak{T}_0$ *if and only if the divisor* $\chi(\mathfrak{Q}_\sigma)$ *of the torsion submodule* $\mathfrak{Q}_\sigma$ *of* C_σ *vanishes*, i. e. the support of $\mathfrak{Q}_\sigma$ has codimension ≥ 2.

Assume $\mathfrak{T}_0$ to be prime. Then $\mathrm{div}\,\mathfrak{R}$ is a multiple of $\mathrm{div}\,\mathfrak{T}_0$. Since $\mathfrak{T}_0 \bar{C}_\sigma = 0$, the length of $(\bar{C}_\sigma)_{\mathfrak{T}_0}$ is simply the rank of $\bar{C}_\sigma$ over the integral domain $A/\mathfrak{T}_0$. Thus:

13.4 Proposition *If the elimination ideal* $\mathfrak{T}_0$ *is a prime ideal in* A,

$$\mathrm{div}\,\mathfrak{R} = (\mathrm{rank}_{A/\mathfrak{T}_0} \bar{C}_\sigma) \cdot \mathfrak{T}_0.$$

In particular, in this situation the equality $\mathfrak{R} = \mathfrak{T}_0$ *holds if and only if* $\bar{C}_\sigma$ *has rank 1 over* $A/\mathfrak{T}_0$.

Let us consider briefly the case that even the ideal $\mathfrak{T}$ of inertia forms is prime, that means that the projective variety defined by the polynomials $F_0, \ldots, F_n$ is irreducible.

13.5 Proposition *If* $\mathfrak{T}$ *is a prime ideal, the following conditions are equivalent:* (1) $\mathfrak{R} = \mathfrak{T}_0$. (2) $\mathrm{rank}_{A/\mathfrak{T}_0} \bar{C}_\sigma = 1$. (3) $\chi(\mathfrak{Q}_\sigma) = 0$. (4) $\mathfrak{Q}_\sigma = 0$. (5) C_σ *is torsionfree.*

Proof. Only the implication $(3) \Rightarrow (4)$ has yet to be proved. $\bar{C}_\sigma = C_\sigma/\mathfrak{T}_\sigma$ is a homogeneous part of the integral domain $C/\mathfrak{T}$ and therefore torsionfree over $A/\mathfrak{T}_0$. The canonical inclusion $\mathfrak{Q}_\sigma \to \bar{C}_\sigma$ shows that $\mathfrak{Q}_\sigma$ is also torsionfree as an $(A/\mathfrak{T}_0)$-module. Therefore $\chi(\mathfrak{Q}_\sigma) = 0$ implies $\mathfrak{Q}_\sigma = 0$. $\square$

In the classical case $\gamma_0 = \cdots = \gamma_n = 1$ condition 5 of Proposition 13.5 implies that C_m is torsionfree even for all $m \leq \sigma$. We show more generally: *If in the classical case C_m is torsionfree then C_{m-1} is torsionfree as well.* But for $q_{m-1} \in \mathfrak{Q}_{m-1}$ we have $q_{m-1} C_1 \subseteq \mathfrak{Q}_m = 0$. This implies $q_{m-1} C_+ = 0$ and hence $q_{m-1} \in \mathfrak{T}_{m-1} \cap \mathfrak{Q}_{m-1} = 0$.

Supplement

1. Let $n = 1$ and let γ, δ be the strictly admissible pairs $(2,3),(6,12)$. Furthermore, let $Q := \mathbb{Z}[X, Y, U, V, W]$.

a) The saturated generic polynomials of degree 6 and 12 over Q are
$$XT_0^3 + YT_1^2, \quad UT_0^6 + VT_0^3 T_1^2 + WT_1^4.$$
Then $\mathfrak{f}_0(\bar{C}_\sigma) = (X(XW - YV) + Y^2 U)$ is prime, $\mathfrak{f}_1(C_\sigma) = (X, Y)$. Thus $\mathfrak{T}_0 = \mathfrak{f}_0(\bar{C}_\sigma) = \mathfrak{R}$ and $\operatorname{div} \mathfrak{f}_1(C_\sigma) = 0$.

b) Let the integrally closed domain $A := k[x, y, z] = k[X, Y, Z]/(Y^2 - XZ)$, k any field, be the specialization of Q given by $X \mapsto x$, $Y \mapsto y$, $U \mapsto 1$, $V \mapsto y$, $W \mapsto z$. The specialized polynomials
$$xT_0^3 + yT_1^2, \quad T_0^6 + yT_0^3 T_1^2 + zT_1^4$$
form a regular sequence in $A[T]$. Then $\mathfrak{R} = (y^2)$. The Fitting ideal $\mathfrak{f}_1(C_\sigma)$ is the prime ideal $\mathfrak{p} := (x, y)$ of codimension 1 which is not principal. Because of $\operatorname{div} y = \mathfrak{p} + \mathfrak{q}$, $\mathfrak{q} := (z, y)$, and $\operatorname{div} \mathfrak{T}_0 = \operatorname{div} \mathfrak{R} - \operatorname{div} \mathfrak{f}_1(C_\sigma) = 2\mathfrak{p} + 2\mathfrak{q} - \mathfrak{p} = \mathfrak{p} + 2\mathfrak{q} = \mathfrak{p} + \operatorname{div} z$ we have $\mathfrak{T}_0 = \mathfrak{p}z$ which is not a principal ideal.

c) Let the integrally closed domain $A := \mathbb{Z}[\sqrt{-5}]$ be the specialization of Q given by $X \mapsto 2$, $Y \mapsto 1 + \sqrt{-5}$, $U \mapsto 1$, $V \mapsto 1 - \sqrt{-5}$, $W \mapsto 3$. The specialized polynomials
$$2T_0^3 + (1 + \sqrt{-5})T_1^2, \quad T_0^6 + (1 - \sqrt{-5})T_0^3 T_1^2 + 3T_1^4$$
form again a regular sequence in $A[T]$. Then $\mathfrak{R} = (2(-2 + \sqrt{-5}))$. The Fitting ideal $\mathfrak{f}_1(C_\sigma)$ is the prime ideal $\mathfrak{p} := (2, 1 + \sqrt{-5})$. Because of $\operatorname{div} 2 \cdot (-2 + \sqrt{-5}) = 2\mathfrak{p} + 2\mathfrak{q}$, $\mathfrak{q} := (3, 1 + \sqrt{-5})$, and $\operatorname{div} \mathfrak{T}_0 = 2\mathfrak{p} + 2\mathfrak{q} - \mathfrak{p} = \mathfrak{p} + 2\mathfrak{q} = \mathfrak{p} + \operatorname{div}(-2 + \sqrt{-5})$ we have $\mathfrak{T}_0 = \mathfrak{p}(-2 + \sqrt{-5})$ which is not a principal ideal.

d) Finally let the integrally closed domain A, defined as a residue class ring by $k[x, y, v, w] = k[X, Y, V, W]/(XW - YV)$, k any field, be the specialization of Q given by $X \mapsto x$, $Y \mapsto y$, $U \mapsto 1$, $V \mapsto v$, $W \mapsto w$. The specialized polynomials
$$xT_0^3 + yT_1^2, \quad T_0^6 + vT_0^3 T_1^2 + wT_1^4$$
form a regular sequence in $A[T]$ with $\mathfrak{R} = (y^2)$. The Fitting ideal $\mathfrak{f}_1(C_\sigma)$ is the divisorial prime ideal $\mathfrak{p} := (x, y)$. Because of $\operatorname{div} \mathfrak{R} = 2\operatorname{div} y = 2\mathfrak{p} + 2\mathfrak{q}$, $\mathfrak{q} := (w, y)$, and $\operatorname{div} \mathfrak{T}_0 = 2\mathfrak{p} + 2\mathfrak{q} - \mathfrak{p} = \mathfrak{p} + 2\mathfrak{q}$, we have $\mathfrak{T}_0 = \mathfrak{q}y$ which is not a principal ideal. (Its divisor class has even infinite order in the class group $\operatorname{Cl} A$.)

14 Resultant Divisors and Duality

We continue with the discussion of the resultant ideal introduced in the last section. As there let A be an integrally closed noetherian domain, $F_0, \ldots, F_n$ a regular sequence of homogeneous polynomials in the polynomial ring $A[T] = A[T_0, \ldots, T_n]$ and $C := A[T]/(F_0, \ldots, F_n)$. We will show now that the resultant ideal $\mathfrak{R} = \mathfrak{R}(F_0 \ldots, F_n)$ describes exactly the divisorial part of the A-algebra of global sections in $\mathcal{O}_{\mathrm{Proj}\,C}$, which is annihilated by the elimination ideal $\mathfrak{T}_0$ and therefore an algebra over $\bar{A} = A/\mathfrak{T}_0 = \bar{C}_0$ in a natural way, cf. [44].

14.1 Theorem *Let A be an integrally closed noetherian domain. For a regular sequence $F_0, \ldots, F_n$ of homogeneous polynomials of positive degrees in $A[T] = A[T_0, \ldots, T_n]$ and $C = A[T]/(F_0, \ldots, F_n)$,*

$$\chi(\Gamma(\mathcal{O}_{\mathrm{Proj}\,C})) = \mathrm{div}\,\mathfrak{R}$$

where $\mathfrak{R}$ is the resultant ideal of the polynomials $F_0, \ldots, F_n$. Explicitly, for every prime ideal $\mathfrak{p}$ of codimension 1 in A we have the equalities

$$\mathrm{length}\,\Gamma(\mathcal{O}_{\mathrm{Proj}\,C})_{\mathfrak{p}} = \mathrm{length}\,\Gamma(\mathcal{O}_{\mathrm{Proj}\,C_{\mathfrak{p}}}) = \mathrm{length}\,(\bar{C}_\sigma)_{\mathfrak{p}} = \mathrm{length}\,(A/\mathfrak{R})_{\mathfrak{p}}.$$

Before proving 14.1 we give a definition and derive some corollaries.

Definition In the situation of Theorem 14.1 the degree

$$\sum_{\mathrm{codim}\,\mathfrak{p}=1} \mathrm{length}\,\Gamma(\mathcal{O}_{\mathrm{Proj}\,C})_{\mathfrak{p}} = \sum_{\mathrm{codim}\,\mathfrak{p}=1} \mathrm{length}(A/\mathfrak{R})_{\mathfrak{p}}$$

of the divisor $\chi(\Gamma(\mathcal{O}_{\mathrm{Proj}\,C})) = \mathrm{div}\,\mathfrak{R}$ is called the (total) d e g r e e o f e l i m i n a t i o n of $\mathrm{Proj}\,C$ over $\mathrm{Spec}\,A$.

If the ideal $\mathfrak{T}$ of inertia forms is prime, $\mathrm{Proj}\,C$ is irreducible and the degree of elimination is nothing else but the degree of the rational function field of $\mathrm{Proj}\,C$ over the function field $Q(A/\mathfrak{T}_0)$ of $\mathrm{Spec}\,A/\mathfrak{T}_0$. An immediate consequence of 14.1 is:

14.2 Corollary *If in addition to the assumptions of Theorem 14.1 the ideal $\mathfrak{T}$ of inertia forms is prime, the degree of elimination equals the rank of $\bar{C}_\sigma = C_\sigma/\mathfrak{T}_\sigma = C_\sigma/A\Delta$ over $A/\mathfrak{T}_0$.*

In particular, the equivalent conditions in Proposition 13.5 are also equivalent to the following condition:

(6) $L = K$, *i. e. the projection $\mathrm{Proj}\,C = \mathrm{Proj}\,C/\mathfrak{T} \to \mathrm{Spec}\,A/\mathfrak{T}_0$ is birational.*

By Theorem 11.10 we have the following conclusion:

14.3 Corollary *For a strictly admissible sequence of generic homogeneous polynomials* $F_j = \sum_{\nu \in N_j} U_{j\nu} T^\nu \in \mathbb{Z}[U_{j\nu}]_{j,\nu}[T]$, $j = 0, \ldots, n$, *with degree of elimination* $d := [\Gamma : (\Gamma_0 + \cdots + \Gamma_n)]$ *we have*

$$\mathfrak{R} = \mathfrak{T}_0^d.$$

Recall that Γ is the group of syzygies of the weights $\gamma_0, \ldots, \gamma_n$ of the indeterminates $T_0, \ldots, T_n$ and Γ_j its subgroup generated by the differences $\nu - \mu$, $\nu, \mu \in N_j$.

For the classical case Corollary 14.3 yields:

14.4 Corollary *In case* $\gamma_0 = \cdots = \gamma_n = 1$ *we have* $\mathfrak{R} = \mathfrak{T}_0$ *for saturated generic homogeneous polynomials* $F_0, \ldots, F_n \in \mathbb{Z}[U_{j\nu}]_{j,\nu}[T]$.

Proof. In this case simply $\Gamma = \Gamma_0 = \cdots = \Gamma_n$.   □

Proof of 14.1. After localizing by $\mathfrak{p}$ we may assume that A is a discrete valuation ring with maximal ideal $\mathfrak{p} = \pi A$. In this case $L := \Gamma(\mathcal{O}_{\mathrm{Proj}\, C})$ is the homogeneous part of degree zero of the total graded quotient ring of $\bar{C} = C/\mathfrak{T}$.

C is a noetherian graded Macaulay ring of dimension 1. This is true for $\bar{C}$, too, provided the unmixed ideal $\mathfrak{T}$ of inertia forms is a proper ideal. This may be assumed, however, because otherwise C is a finite complete intersection over A and $\mathrm{Proj}\, C = \emptyset$, $A = \mathfrak{T}_0 = \mathfrak{R}$.

Trivially, $\mathrm{length}\, L \geq \mathrm{length}\, \bar{C}_r$ for every $r \in N_C$, where

$$N_C$$

denotes the set of $r \in \mathbb{N}$ such that $\bar{C}_r$ contains a non-zero-divisor of $\bar{C}$. The subset N_C of $\mathbb{N}$ is a non-zero submonoid of $\mathbb{N}$. It is easily seen that there is a $r_0 \in N_C$ such that $\mathrm{length}\, L = \mathrm{length}\, \bar{C}_r$ for every $r \in N_C$, $r \geq r_0$.

To finish, we just have to apply a conclusion drawn from the following lemma, namely Corollary 14.6 below.   □

14.5 Lemma *Assume A to be a discrete valuation ring. Let $\mathfrak{T}_0 \neq A$ and $\bar{A} := A/\mathfrak{T}_0 = \bar{C}_0$. For every homogeneous element $z \in C$ of positive degree r, which is a non-zero-divisor in $\bar{C}$, there is an isomorphism*

$$\mathfrak{Q}/z\mathfrak{Q} \cong \mathrm{Hom}_{\bar{A}}(\bar{C}/z\bar{C}, \bar{A})(-\sigma - r)$$

of graded C-modules.

$\mathfrak{Q}$ denotes as usual the irrelevant component of the zero ideal in C. Recall that for a graded module V and $m \in \mathbb{Z}$, the graded module $V(m)$ has $V(m)_i := V_{m+i}$ as homogeneous part of degree i.

14.6 Corollary *In the situation of* 14.5, *for every* $r \in N_C$ *and* $s \in \mathbb{Z}$

$$\text{length}_A \mathfrak{Q}_{r+s} - \text{length}_A \mathfrak{Q}_s = \text{length}_A \bar{C}_{\sigma - s} - \text{length}_A \bar{C}_{\sigma - s - r}.$$

In particular $(s = 0)$, *for every* $r \in N_C$ *with* $r > \sigma$

$$\text{length}_A \bar{C}_\sigma = \text{length}_A \mathfrak{Q}_r \ (= \text{length}_A C_r = \text{length}_A \bar{C}_r).$$

Proof. The corollary follows from 14.5 simply by the fact that dualizing by $\bar{A}$ preserves lengths. □

In general, σ does not belong to N_C, cf. Supplement 1.

Proof of Lemma 14.5. Let $A[Z]$ denote the graded polynomial algebra over A in the single indeterminate Z of degree r. Then C and $\bar{C}$ are finite graded $A[Z]$-algebras by $Z \mapsto z$ and

$$\text{Hom}_{\bar{A}}(\bar{C}/z\bar{C}, \bar{A}) = \text{Hom}_{\bar{A}[Z]}(\bar{C}/z\bar{C}, \bar{A}) = \text{Hom}_{\bar{A}[Z]}(\bar{C}, \bar{A}).$$

Corresponding to the exact sequence $0 \to \bar{A}[Z] \xrightarrow{Z} \bar{A}[Z] \to \bar{A} \to 0$ there is the exact sequence

$$0 \longrightarrow \text{Hom}_{\bar{A}[Z]}(\bar{C}, \bar{A}[Z]) \xrightarrow{Z} \text{Hom}_{\bar{A}[Z]}(\bar{C}, \bar{A}[Z])$$
$$\longrightarrow \text{Hom}_{\bar{A}[Z]}(\bar{C}, \bar{A}) \longrightarrow \text{Ext}^1_{\bar{A}[Z]}(\bar{C})$$

where however $\text{Ext}^1_{\bar{A}[Z]}(\bar{C}) := \text{Ext}^1_{\bar{A}[Z]}(\bar{C}, \bar{A}[Z])$ vanishes, which is proved as follows: From the exact sequence $0 \to \bar{C} \xrightarrow{z} \bar{C} \to \bar{C}/z\bar{C} \to 0$ we get an exact sequence

$$\text{Ext}^1_{\bar{A}[Z]}(\bar{C}) \xrightarrow{z} \text{Ext}^1_{\bar{A}[Z]}(\bar{C}) \longrightarrow \text{Ext}^2_{\bar{A}[Z]}(\bar{C}/z\bar{C}, \bar{A}[Z]).$$

The Ext^2-group is zero because $\bar{A}[Z]$ is a Gorenstein ring of dimension 1. Hence multiplication by z on the finite graded $\bar{A}[Z]$-module $\text{Ext}^1_{\bar{A}[Z]}(\bar{C})$ is surjective, which implies that the module vanishes.

Here, as during the rest of the proof, one has to observe that all modules $\text{Ext}(V, W)$ considered are graded modules in a natural way because the modules V used therein happen to be finite modules, such that the extension groups coincide with those constructed within the category of graded noetherian rings and graded modules, cf. [6], Section 1.5.

It suffices to show now that

$$\text{Hom}_{\bar{A}[Z]}(\bar{C}, \bar{A}[Z]) = \text{Hom}_{A[Z]}(\bar{C}, \bar{A}[Z]) \cong \mathfrak{Q}(\sigma + r)$$

as graded C-modules.

First we prove that the homomorphism $C \to \bar{C}$ induces an isomorphism

$$\text{Hom}_{A[Z]}(\bar{C}, \bar{A}[Z]) = \text{Hom}_{A[Z]}(C, \bar{A}[Z])$$

by showing that $\mathrm{Hom}_{A[Z]}(\mathfrak{T}, \bar{A}[Z]) = 0$. We consider the exact sequence

$$\mathrm{Hom}_{A[Z]}(\mathfrak{T}, A[Z]) \longrightarrow \mathrm{Hom}_{A[Z]}(\mathfrak{T}, \bar{A}[Z])$$
$$\longrightarrow \mathrm{Ext}^1_{A[Z]}(\mathfrak{T}, A[Z]) \xrightarrow{\pi^s} \mathrm{Ext}^1_{A[Z]}(\mathfrak{T}, A[Z])$$

belonging to $0 \to A[Z] \xrightarrow{\pi^s} A[Z] \to \bar{A}[Z] \to 0$, where π^s is a generator of $\mathfrak{T}_0$. For trivial reasons, $\mathrm{Hom}_{A[Z]}(\mathfrak{T}, A[Z]) = 0$. Therefore it remains to show that $\mathrm{Ext}^1_{A[Z]}(\mathfrak{T}, A[Z])$ is a torsionfree A-module. But this follows from an isomorphism

$$\mathrm{Ext}^1_{A[Z]}(\mathfrak{T}, A[Z]) \;=\; \mathrm{Hom}_A(\mathfrak{T}, A)$$

of modules. For the construction of this canonical isomorphism see [39], Lemma 4.1; one has to use that $\mathfrak{T}$ is a finite free A-module. A modified version of the general construction is given in Supplement 6 for the reader's convenience. In the special situation considered here the proof is elementary, see Supplement 5.

Considering multiplication by π^s again, we get an exact sequence

$$\mathrm{Hom}_{A[Z]}(C, A[Z]) \longrightarrow \mathrm{Hom}_{A[Z]}(C, \bar{A}[Z])$$
$$\longrightarrow \mathrm{Ext}^1_{A[Z]}(C, A[Z]) \xrightarrow{\pi^s} \mathrm{Ext}^1_{A[Z]}(C, A[Z]) \,.$$

Again, $\mathrm{Hom}_{A[Z]}(C, A[Z]) = 0$ for trivial reasons. Thus $\mathrm{Hom}_{A[Z]}(C, \bar{A}[Z])$ is nothing else but the π^s-torsion of $\mathrm{Ext}^1_{A[Z]}(C, A[Z])$. Because $\mathfrak{Q}$ is the π^s-torsion of C, it finally suffices to show that there is an isomorphism

$$\mathrm{Ext}^1_{A[Z]}(C, A[Z]) \;\cong\; C(\sigma + r)$$

of graded C-modules.

C has the following representation:

$$C \;=\; C[Z]/(Z - z) \;=\; A[Z, T]/(F_0, \ldots, F_n, Z - z)$$

where $F_0, \ldots, F_n, Z - z$ is a regular sequence in $A[Z, T]$. Furthermore, C has homological dimension 1 over $A[Z]$ because C is finite over $A[Z]$ and is a Macaulay ring of dimension 1. Thus there is a canonical isomorphism

$$\mathrm{Ext}^1_{A[Z]}(C, A[Z]) \;\cong\; \mathrm{Ext}^{n+2}_{A[Z,T]}(C, A[Z, T])(-\gamma_0 - \cdots - \gamma_n)$$

(γ_i being the degree of T_i) of graded C-modules, cf. again [39], Lemma 4.1, or the short treatment in Supplement 6. From the Koszul resolution of C over $A[Z, T]$ one gets the isomorphism of graded C-modules:

$$\mathrm{Ext}^{n+2}_{A[Z,T]}(C, A[Z, T]) \;\cong\; C(\delta_0 + \cdots + \delta_n + r)$$

(δ_i being the degree of F_i, r the degree of $Z - z$). The composition of both isomorphisms yields an isomorphism $\mathrm{Ext}^1_{A[Z]}(C, A[Z]) \cong C(\sigma + r)$ we were looking for because of $-\gamma_0 - \cdots - \gamma_n + \delta_0 + \cdots + \delta_n + r = \sigma + r$. □

We finish this section giving some applications of Corollary 14.6. First we will show that the resultant ideal $\mathfrak{R} = \mathfrak{d}(\bar{C}_\sigma)$ with $\operatorname{div}\mathfrak{R} = \operatorname{div}\mathfrak{f}_0(\bar{C}_\sigma) = \chi(\bar{C}_\sigma)$ can be computed from some other homogeneous parts $C_r = \bar{C}_r$, $r > \sigma$, of C, too. To formulate this we denote by

$$N_C^u$$

the set of $r \in \mathbb{N}$ such that C_r is not contained in any prime ideal $\mathfrak{P} \in \operatorname{Ass}_C \bar{C}$ with $\operatorname{ht}_A \mathfrak{P}_0 = 1$. Note that N_C^u contains the set N_C of those integers $r \in \mathbb{N}$ for which $\bar{C}_r$ contains a non-zero-divisor of $\bar{C}$. (This set N_C was introduced earlier in the proof of 14.1.)

14.7 Theorem *Let A be an integrally closed noetherian domain. For a regular sequence $F_0, \ldots, F_n$ of homogeneous polynomials of positive degrees in $A[T] = A[T_0, \ldots, T_n]$, $C = A[T]/(F_0, \ldots, F_n)$ and every $r \in N_C^u$, $r > \sigma$,*

$$\operatorname{div}\mathfrak{R} = \operatorname{div}\mathfrak{f}_0(C_r) = \chi(C_r).$$

Proof. We have to show that $\operatorname{length}(C_r)_\mathfrak{p} = \operatorname{length} A_\mathfrak{p}/\mathfrak{R}_\mathfrak{p} = \operatorname{length}(\bar{C}_\sigma)_\mathfrak{p}$ for every prime ideal $\mathfrak{p} \subseteq A$ of codimension 1. For this we may assume that $A = A_\mathfrak{p}$ is a discrete valuation ring. Using a Kronecker extension of A we may also assume that the residue field of A is infinite. In this case $C_r = \bar{C}_r$ contains a non-zero-divisor of $\bar{C}$, cf. Supplement 3. By 14.6, $\operatorname{length}\bar{C}_\sigma = \operatorname{length} C_r$.   □

Any common multiple r of the weights $\gamma_0, \ldots, \gamma_n$ fulfills the condition which had to be presumed in Theorem 14.7, cf. Supplement 3a). In particular, in the classical case $\gamma_0 = \cdots = \gamma_n = 1$ every $r > \sigma$ can be used to compute the resultant ideal. For $n = 1$ and $r = \sigma + 1$ one recovers the classical representation of the resultant as the Sylvester determinant. For arbitrary n Macaulay used the link of the resultant ideal with the Fitting ideal $\mathfrak{f}_0(C_{\sigma+1})$ to define the resultant ideal of generic polynomials, see [30], Section 1. See also the article [8] by M. Chardin. (In [8] Koszul complexes are used extensively, too.)

In this context we mention the following result:

14.8 Corollary *Let A be an integrally closed noetherian domain and let $F_0, \ldots, F_n$ be a regular sequence of homogeneous polynomials of positive degrees $\delta_0, \ldots, \delta_n$ in $A[T] = A[T_0, \ldots, T_n]$. Assume that the pair $(\gamma_0, \ldots, \gamma_n)$, $(\delta_0, \ldots, \delta_n)$ is strictly admissible and that $\gcd(\gamma_0, \ldots, \gamma_n) = 1$. Then for $C = A[T]/(F_0, \ldots, F_n)$ and every $r > \sigma$*

$$\operatorname{div}\mathfrak{R} = \operatorname{div}\mathfrak{f}_0(C_r) = \chi(C_r).$$

Proof. First we consider the case that the polynomials $F_j \in \mathbb{Z}[U_{j\nu}]_{j,\nu}[T]$ are generic. Let $r > \sigma$. By 14.7 it suffices to show that $C_r = \bar{C}_r$ contains a non-zero-divisor of $\bar{C}$. By definition of strict admissibility, cf. Section 8, the residue classes of all indeterminates and hence of all monomials in these indeterminates are non-zero-divisors in $\bar{C}$. The assertion follows now from the fact that $r \in \mathrm{Mon}(\gamma_0, \ldots, \gamma_n)$ which is a special case of part (2) of the next theorem.

In the general case the assertion is obtained from the generic situation by a general result on change of base rings which will be proved in Supplement 4 of Section 15.   □

If $d := \gcd(\gamma_0, \ldots, \gamma_n) > 1$, then the statement of Corollary 14.8 is true for all $r > \sigma$ which are multiples of d. (Note: σ itself is a multiple of d.)

For further similar applications of 14.7 we refer to Supplement 5 of Section 15.

In the next application some inequalities will be proved, which relate properties of strictly admissible pairs γ, δ to structural constants of the numerical monoid $\mathrm{Mon}(\gamma)$, cf. Section 3.

14.9 Theorem *Let γ, δ be a strictly admissible pair of weights and positive degrees. Then the following holds:*
(1) $\mathrm{sat}_\gamma \leq g_\gamma + \max_{j \neq i} \delta_j$ *for* $i = 0, \ldots, n$.
(1′) $\mathrm{sat}_\gamma \leq g_\gamma + \max_j \delta_j$.
(2) $g_\gamma + \max_j \delta_j \leq \sigma$.
(3) $\mathrm{vsat}_\gamma \leq \sigma$.

Proof. To prove (1), consider $m \in \mathbb{N}$ with $m > g_\gamma + \delta_j$ for $j \neq i$. Then $m - \delta_j \in \mathrm{Mon}(\gamma)$. Therefore $\Gamma_j = \Gamma(\delta_j) \subseteq \Gamma(m)$, $j \neq i$. By Theorem 11.10(1) the index $[\Gamma : \Gamma(m)]$ is finite and $m \in \mathrm{Sat}(\gamma)$. (1′) is a consequence of (1).

To prove (2) and (3), we consider the generic polynomials
$$F_j = \sum\nolimits_{\nu \in \mathbb{N}_{\delta_j}(\gamma)} U_{j\nu} T^\nu \in Q[T], \quad j = 0, \ldots, n,$$
and $C = Q[T]/(F_0, \ldots, F_n)$. The Poincaré series of C is the polynomial
$$\mathcal{P}_C = \frac{(1 - Z^{\delta_0}) \cdots (1 - Z^{\delta_n})}{(1 - Z^{\gamma_0}) \cdots (1 - Z^{\gamma_n})}$$
of degree σ. To the algebra $C_{\langle j \rangle} := Q[T]/(F_0, \ldots, \hat{F}_j, \ldots, F_n)$ belongs the Poincaré series $\mathcal{P}_j = \mathcal{P}_C/(1 - Z^{\delta_j})$. By Supplement 7 of Section 11, $\mathcal{P}_j$ gets stationary at $\sigma + 1 - \delta_j$ (with a positive limit of coefficients). This implies $(C_{\langle j \rangle})_m \neq 0$ for $m \geq \sigma + 1 - \delta_j$ and thus $Q[T]_m \neq 0$, which means $m \in \mathrm{Mon}(\gamma)$. This proves (2).

To prove (3), consider $m \in \mathbb{N}$ with $m > \sigma$. By (2), $m - \delta_j \in \mathrm{Mon}(\gamma)$. Therefore, $\Gamma(m) \supseteq \Gamma_j = \Gamma(\delta_j)$.

The ideal $\mathfrak{T} \subseteq C$ of inertia forms is prime since the pair γ, δ is strictly admissible. The residue classes t_i of T_i are not in $\mathfrak{T}$, because $\mathfrak{T}$ coincides with the principal component, cf. Lemma 8.8. Thus the monoid of those $r \in \mathbb{N}$, for which C_r is not contained in $\mathfrak{T}$, is exactly $\mathrm{Mon}(\gamma)$. Because $m \in \mathrm{Mon}(\gamma)$, $\chi(C_m) = \mathrm{div}\,\mathfrak{R}$ by Theorem 14.7. This means

$$\mathrm{rank}_{Q/\mathfrak{T}_0}C_m = [L:K] = d = [\Gamma : (\Gamma_0 + \cdots + \Gamma_n)],$$

where L is the function field of $\mathrm{Proj}\,C$ and where K is the quotient field of $Q/\mathfrak{T}_0$, cf. 14.3. As a consequence, there are monomials $t^{\nu_1}, \ldots, t^{\nu_d} \in C_m$ such that $t^0, t^{\nu_2-\nu_1}, \ldots, t^{\nu_d-\nu_1}$ is a K-basis of L. By Theorem 11.10(2) there is a subfield $K^\flat$ of K such that $K = K^\flat(t^{\beta(1)}, \ldots, t^{\beta(n)})$, $L = K^\flat(t^{\alpha(1)}, \ldots, t^{\alpha(n)})$ where $\beta(1), \ldots, \beta(n)$ is a $\mathbb{Z}$-basis of $\Gamma_0 + \cdots + \Gamma_n$ and $\alpha(1), \ldots, \alpha(n)$ is a $\mathbb{Z}$-basis of Γ and where $t^{\beta(1)}, \ldots, t^{\beta(n)}$ (as well as $t^{\alpha(1)}, \ldots, t^{\alpha(n)}$) are algebraically independent over $K^\flat$. Now consider the finite extensions

$$K^\flat[\Gamma_0 + \ldots + \Gamma_n] \subseteq K^\flat[\Gamma'] \subseteq K^\flat[\Gamma]$$

of group algebras, $\Gamma' := \Gamma_0 + \ldots + \Gamma_n + \sum_{k=1}^d \mathbb{Z}(\nu_k - \nu_1) \subseteq \Gamma(m)$. Because the quotient fields of $K^\flat[\Gamma']$ and $K^\flat[\Gamma]$ coincide, $\Gamma' = \Gamma$ and therefore $\Gamma(m) = \Gamma$. $\qquad\square$

The inequalities in Theorem 14.9 may actually be equalities, cf. Supplement 8. Furthermore, the inequality $\mathrm{vsat}_\gamma \leq \mathrm{g}_\gamma + \max_j \delta_j$ may not be true, cf. Supplement 9.

Supplements

1. Let A be a discrete valuation ring with maximal ideal πA, $n = 2$, $\gamma = (3,4,5)$ and $C = A[T]/(F_0, \ldots, F_n)$. Compute $\mathfrak{T}$, $\sqrt{\mathfrak{T}}$, N_C, $\mathfrak{Q}$ and $\mathfrak{R}$, $\mathfrak{T}_0$, C_σ, $\mathfrak{Q}_\sigma$, $\bar{C}_\sigma$ in the following cases (where X, Y, Z stand for T_0, T_1, T_2, cf. also Lemma 14.5 and its proof).
a) $F_0 = XZ + \pi Y^2$, $F_1 = \pi X^3 + YZ$, $F_2 = X^2Y + \pi Z^2$.
b) $F_0 = XZ + \pi Y^2$, $F_1 = X^3 + \pi YZ$, $F_2 = \pi X^2Y + Z^2$.
c) $F_0 = \pi XZ + Y^2$, $F_1 = X^3 + \pi YZ$, $F_2 = \pi X^2Y + Z^2$.
c) $F_0 = \pi XZ + Y^2$, $F_1 = \pi X^3 + YZ$, $F_2 = \pi X^2Y + Z^2$.
  In the following example let γ be $(2,3,5)$:
e) $F_0 = X^2$, $F_1 = Y^2$, $F_2 = XY + \pi Z$.

2. Let A be a discrete valuation ring and $F_0, \ldots, F_n$ a regular sequence in $A[T]$ of homogeneous polynomials; adopt the usual conventions about C, $\mathfrak{Q}$, $\mathfrak{T}$, $\bar{C}$ etc. Let $z \in C_r$ be a homogeneous element of positive degree r. Then z is a non-zero-divisor in $\bar{C}$ if and only if C/zC is a finite A-algebra.
  If this is the case, $(C/zC)_m = 0$ for every $m \geq \sigma + r$.

3. Let A be a discrete valuation ring and $F_0, \ldots, F_n$ a regular sequence in $A[T]$ of homogeneous polynomials; adopt the usual conventions about C, $\mathfrak{Q}$, $\mathfrak{T}$, $\bar{C}$, N_C, N_C^u etc.

a) N_C^u is a submonoid of $\mathbb{N}$ containing N_C. The least common multiple of $\gamma_0, \ldots, \gamma_n$ belongs to N_C^u.
b) If the residue class field of A is not finite, $N_C^u = N_C$. (One can always switch to this situation by applying a (faithfully flat) Kronecker extension of the ground ring A.) In any case, $N_C^u \setminus N_C$ is a finite set.
c) Corollary 14.6 holds for every $r \in N_C^u$ and $s \in \mathbb{Z}$.
d) In the classical case $\gamma_0 = \cdots = \gamma_n = 1$, $N_C^u = \mathbb{N}$. Thus in this case

$$\text{length}_A C_r = \text{length}_A \bar{C}_\sigma$$

for every $r > \sigma$. (If $N_C^u = N_C$, C_r and $\bar{C}_\sigma$ are even isomorphic.)

4.† The proposition given in Suppl. 5 of Sect. 12 is a simple consequence of 14.6. (Apply 14.6 with $s = m - r$, $s = \sigma - m$ and $r > m$.)

5. Let V be a finite projective A-module and $\alpha \in \text{End}_A V$. Then V is a module over $A[X]$ by $Xv = \alpha(v)$, $v \in V$. There is a canonical $A[X]$-operator Φ_α on $A[X] \otimes_A V$ such that $\Phi_\alpha(f \otimes v) = Xf \otimes v - f \otimes Xv$, $f \in A[X]$, $v \in V$, and a canonical exact $A[X]$-sequence

$$0 \longrightarrow A[X] \otimes_A V \xrightarrow{\Phi_\alpha} A[X] \otimes_A V \longrightarrow V \longrightarrow 0.$$

Let β denote the dual operator of α on $\text{Hom}_A(V, A)$. The canonical isomorphism

$$t_V \ : \ A[X] \otimes_A \text{Hom}_A(V, A) \longrightarrow \text{Hom}_{A[X]}(A[X] \otimes_A V, A[X])$$

over $A[X]$ transports Φ_β into $\text{Hom}(\Phi_\alpha, A[X])$. As a result, there is an induced $A[X]$-isomorphism

$$\text{Hom}_A(V, A) \longrightarrow \text{Ext}^1_{A[X]}(V, A[X]).$$

Assume in addition that A is a graded ring, V a graded finite projective A-module and that α is homogeneous of degree $m \in \mathbb{Z}$. The polynomial algebra $A[X]$ has a natural grading with $\deg X = m$. Then Φ_α is a homogeneous operator on the graded $A[X]$-module $A[X] \otimes_A V$ of degree m and Φ_β is homogeneous of degree m, too. The canonical isomorphism t_V is homogeneous of degree 0. From this results an isomorphism of graded $A[X]$-modules

$$\text{Hom}_A(V, A) \longrightarrow \text{Ext}^1_{A[X]}(V, A[X])(-m).$$

6. The result of Suppl. 5 can be generalized in the following way: Let V be a finite A-module of homological dimension $\leq r$ and $\alpha \in \text{End}_A V$. Then there is a canonical $A[X]$-isomorphism

$$\text{Ext}^r_A(V, A) \longrightarrow \text{Ext}^{r+1}_{A[X]}(V, A[X]).$$

(For the proof let $0 \to P_r \to \cdots \to P_0 \to V \to 0$ be a resolution of V by finite projective A-modules $P_0, \ldots, P_r$. These can be given a $A[X]$-module structure defined by A-module endomorphisms $\alpha_i : P_i \to P_i$, $i = 0, \ldots, r$, such that the resolution actually is a sequence of modules and homomorphisms over $A[X]$. This means that the α_i together with α define an endomorphism of the resolution. (The following constructions do not really depend on the choice of the α_i because the endomorphism of the resolution given by α and any other $\alpha_i' : P_i \to P_i$ is homotopic to the endomorphism defined by α and the α_i.)

The sequence $0 \to A[X] \otimes_A P_r \to \cdots \to A[X] \otimes_A P_0 \to A[X] \otimes_A V \to 0$ is a resolution of $A[X] \otimes_A V$ by projective $A[X]$-modules. This resolution has an $A[X]$-endomorphism given by the canonical endomorphisms Φ_α and $\Phi_i := \Phi_{\alpha_i}$, $i = 0, \ldots, r$, cf. Suppl. 5 for the definition of these operators. The cokernel simply is our original resolution of V. By $*$ we denote $A[X]$-duals. The cokernels of the

Φ_i^* are $\operatorname{Ext}^1_{A[X]}(P_i, A[X]) = \operatorname{Hom}_A(P_i, A)$, see Suppl. 5. From the commutative diagram

$$
\begin{array}{ccccccc}
(A[X] \otimes_A P_{r-1})^* & \xrightarrow{\ \Phi_{r-1}^*\ } & (A[X] \otimes_A P_{r-1})^* & \longrightarrow & \operatorname{Hom}_A(P_{r-1}, A) & \to & 0 \\
\downarrow & & \downarrow & & \downarrow & & \\
(A[X] \otimes_A P_r)^* & \xrightarrow{\ \Phi_r^*\ } & (A[X] \otimes_A P_r)^* & \longrightarrow & \operatorname{Hom}_A(P_r, A) & \to & 0
\end{array}
$$

we get an exact sequence of cokernels

$$\operatorname{Ext}^r_{A[X]}(A[X] \otimes_A V, A[X]) \to \operatorname{Ext}^r_{A[X]}(A[X] \otimes_A V, A[X]) \to \operatorname{Ext}^r_A(V, A) \to 0$$

where the first homomorphism is $\operatorname{Ext}^r(\Phi_\alpha)$. But the cokernel of $\operatorname{Ext}^r(\Phi_\alpha)$ is $\operatorname{Ext}^{r+1}_{A[X]}(A[X] \otimes_A V, A[X])$.)

In the graded case there is a canonical isomorphism

$$\operatorname{Ext}^r_A(V, A) \longrightarrow \operatorname{Ext}^{r+1}_{A[X]}(V, A[X])(-m)$$

if $\alpha : V \to V$ is homogeneous of degree $m \in \mathbb{Z}$. (See Suppl. 5.)

There is an obvious extension to the case of several variables: Let V be a graded module over the graded polynomial algebra $R = A[X_1, \ldots, X_n]$ with the graded base ring A and homogeneous indeterminates X_i of degree $m_i \in \mathbb{Z}$. Assume that V is finite over A and has homological dimension $\leq r$ over A. Then there is a canonical isomorphism

$$\operatorname{Ext}^r_A(V, A) \longrightarrow \operatorname{Ext}^{r+n}_R(V, R)(-m_1 - \cdots - m_n)$$

of graded R-modules. (Proof by induction on n using the fact that V has homological dimension $\leq r + n - 1$ over $A[X_1, \ldots, X_{n-1}]$.)

7.† Let A be a discrete valuation ring and $F_0, \ldots, F_n$ a regular sequence in $A[T]$ of homogeneous polynomials; adopt the usual conventions about C, $\mathfrak{Q}$, $\mathfrak{T}$, $\bar{C}$, $\bar{A}$. Let $z \in C_r$ be a homogeneous element of positive degree r. Then there is a natural $\bar{A}$-bilinear homogeneous form

$$\mathfrak{Q}/z\mathfrak{Q} \times (\bar{C}/z\bar{C})(\sigma + r) \longrightarrow \bar{A}$$

which defines a perfect duality. (Note that $\mathfrak{Q}/z\mathfrak{Q}$ and $\bar{C}/z\bar{C}$ are finite graded $\bar{A}$-modules. Let $\psi \in C_\sigma^* \cong A$ be a generator. The value of the bilinear form on $([q], [c])$, where $m \in \mathbb{N}$, $q \in \mathfrak{Q}_m$ and $c \in C_{\sigma+r-m}$, can be constructed as

$$\langle [q], [c] \rangle := \overline{\psi(\tfrac{qc}{z})} \in \bar{A}$$

because $qc \in C_{\sigma+r}$ can be written as $qc = zu$ with $u \in C_\sigma$, cf. Suppl. 2, such that $\psi(u) \in A$ and its residue class $\overline{\psi(u)}$ in $\bar{A}$ are well-defined. $\overline{\psi(u)}$ does not depend on the choice of u and depends only on the residue classes $[q] \in \mathfrak{Q}/z\mathfrak{Q}$, $[c] \in \bar{C}/z\bar{C}$. One has to prove, that the bilinear form over $\bar{A}$ is non-degenerate. By Lemma 14.5 there is a homogeneous isomorphism

$$\varphi = \bigoplus_m \varphi_m \ : \ (\mathfrak{Q}/z\mathfrak{Q}) \longrightarrow \operatorname{Hom}_{\bar{A}}((\bar{C}/z\bar{C})(\sigma + r), \bar{A})$$

of graded C-modules. Thus $\varphi_{m+i} \circ \lambda_d = \lambda'_d \circ \varphi_m$ for every homogeneous element $d \in C_i$, where λ_d denotes multiplication by d and λ'_d its $\bar{A}$-dual.)

The proof shows that the perfect duality given above can be written canonically as a perfect duality

$$\mathfrak{Q}/z\mathfrak{Q} \times (\bar{C}/z\bar{C})(\sigma + r) \longrightarrow \mathfrak{T}_0^{-1}/A \ (\cong \bar{A}).$$

In this form the perfect duality holds for arbitrary Dedekind domains A, cf. [44], Sect. 3.

8. Let $a_0, \dots, a_n$ be pairwise relatively prime positive integers, $b := a_0 \cdots a_n$ their product, $\gamma := (\gamma_0, \dots, \gamma_n)$ with $\gamma_i := b/a_i$ and $\delta := (\delta_0, \dots, \delta_n)$ the constant tuple with $\delta_0 = \cdots = \delta_n := b$. Then the pair γ, δ is strictly admissible and

$$\mathrm{sat}_\gamma \;=\; \mathrm{vsat}_\gamma \;=\; \mathrm{g}_\gamma + b \;=\; \sigma \;=\; (n+1)b - (\gamma_0 + \cdots + \gamma_n).$$

(Sect. 3, Suppl. 12.)

9. The pair

$$\gamma := (30, 30, 42, 42, 70, 70, 105, 71), \quad \delta := (142, 142, 210, 210, 210, 210, 210, 210)$$

is strictly admissible. We have $\mathrm{g}_\gamma = 229$, $\mathrm{sat}_\gamma = 439$, $\mathrm{vsat}_\gamma = 454$ and

$$\mathrm{sat}_\gamma \;=\; \mathrm{g}_\gamma + \max_j \delta_j \;<\; \mathrm{vsat}_\gamma .$$

($\mathrm{g}_\gamma = \mathrm{g}_{(30,42,70,105,71)}$ is to be calculated directly. Then $\mathrm{sat}_\gamma \leq \mathrm{g}_\gamma + \max_j \delta_j = 439$ by 14.9. But $439 \notin \mathrm{Sat}(\gamma)$. $454 \notin \mathrm{VSat}(\gamma)$ follows from the fact that the last component in all $\alpha \in \mathrm{N}_{454}(\gamma)$ is 0 or 2. Thus $\mathrm{vsat}_\gamma \geq 454$.

If one replaces $\gamma_7 = 71$ by 101 and $\delta_0 = \delta_1 = 142$ by 202 then $463 = \mathrm{sat}_\gamma = \mathrm{g}_\gamma + 210 < \mathrm{vsat}_\gamma = 484$.)

Note that in the foregoing examples $\Gamma_0 + \cdots + \Gamma_7 \neq \Gamma = \mathrm{Syz}(\gamma)$, i.e. the degree of elimination is > 1. There is an $m > \mathrm{g}_\gamma + \max_j \delta_j$ with $\Gamma(m; \gamma) \neq \Gamma$. But $\Gamma_j \subseteq \Gamma(m; \gamma)$ because of $m - \delta_j \in \mathrm{Mon}(\gamma)$.

For other examples in smaller dimensions see Sect. 11, Suppl. 10.)

15 Resultants

As in Sections 13 and 14 throughout, let A be an integrally closed noetherian domain. We consider a regular sequence $F_0, \ldots, F_n$ of homogeneous polynomials of positive degrees $\delta_0, \ldots, \delta_n$ in the graded polynomial algebra $A[T] = A[T_0, \ldots, T_n]$ with $\deg T_i = \gamma_i > 0$ and the graded A-algebra $C = A[T]/(F_0, \ldots, F_n)$.

Using the finite free A-resolution

$$0 \longrightarrow V_{n+1} \longrightarrow V_n \longrightarrow \cdots \longrightarrow V_1 \longrightarrow V_0 \overset{\varepsilon}{\longrightarrow} C_\sigma \longrightarrow 0$$

of C_σ, which is taken to be the homogeneous part of degree σ of the Koszul resolution $K(A[T]; F_0, \ldots, F_n)$ of C, we are going to construct a canonical generator of the resultant ideal

$$\mathfrak{R} = \mathfrak{R}(F_0, \ldots, F_n) = \mathfrak{R}_A(F_0, \ldots, F_n)$$

which is the principal ideal belonging to the principal divisor $\chi(\bar{C}_\sigma)$, cf. Theorem 13.3. *This canonical generator of $\mathfrak{R}$ we propose to call the* resultant

$$R = \mathrm{R}(F_0, \ldots, F_n) = \mathrm{R}_A(F_0, \ldots, F_n)$$

of the polynomials $F_0, \ldots, F_n$.

Let $h \in A$ be any element having the same zero set as $\mathfrak{T}_0$ and B the ring of fractions A_h (embedded in the quotient field of A and in this way independent of the choice of h). Because of $B \otimes C_\sigma = (C_\sigma)_h \cong B$, cf. 12.5, the sequence

$$0 \overset{f_{n+2}}{\longrightarrow} W_{n+1} \overset{f_{n+1}}{\longrightarrow} W_n \overset{f_n}{\longrightarrow} \cdots \overset{f_2}{\longrightarrow} W_1 \overset{f_1}{\longrightarrow} W_0 \overset{f_0}{\longrightarrow} B \otimes C_\sigma \overset{f_{-1}}{\longrightarrow} 0$$

with $W_i := B \otimes V_i$ and $f_0 := B \otimes \varepsilon$ is a splitting exact sequence of free B-modules. This gives rise to a uniquely determined B-isomorphism

$$B \otimes C_\sigma \otimes \det V_1 \otimes \det V_3 \otimes \cdots \overset{\Theta}{\longrightarrow} B \otimes \det V_0 \otimes \det V_2 \otimes \cdots$$

defined in the following way: Let $g_{-2}, g_{-1}, g_0, \ldots, g_n, g_{n+1}$ be homomorphisms such that the sequence

$$0 \overset{g_{n+1}}{\longleftarrow} W_{n+1} \overset{g_n}{\longleftarrow} W_n \overset{g_{n-1}}{\longleftarrow} \cdots \overset{g_1}{\longleftarrow} W_1 \overset{g_0}{\longleftarrow} W_0 \overset{g_{-1}}{\longleftarrow} B \otimes C_\sigma \overset{g_{-2}}{\longleftarrow} 0$$

is exact and that for all $i = -1, \ldots, n+1$ the identity

$$f_{i+1} g_i + g_{i-1} f_i = \mathrm{id}_{W_i}$$

holds, where W_{-1} stands for $B \otimes C_\sigma$. Let

$$W := \bigoplus_{i=-1}^{n+1} W_i = U \oplus G, \quad U := \bigoplus_{i \text{ odd}} W_i, \quad G := \bigoplus_{i \text{ even}} W_i.$$

Then $\Phi := f + g = \bigoplus_i f_i + \bigoplus_i g_i$ is an involution of W and maps U isomorphically onto G (with inverse $\Phi|G$). The determinant

$$\det(\Phi|U) \ : \ \det U \longrightarrow \det G$$

is independent of the choice of the homomorphisms g_i (cf. Supplement 1) and thus determines the isomorphism Θ from $\det U = \bigotimes_{i \, \text{odd}} \det W_i$ onto $\det G = \bigotimes_{i \, \text{even}} \det W_i$ we were looking for.

The A-modules $V_0, \ldots, V_{n+1}$ are direct sums of homogeneous parts of the polynomial algebra $A[T]$ and therefore have monomial bases in the $T_0, \ldots, T_n$. They determine, uniquely up to sign, bases of

$$\det V_1 \otimes \det V_3 \otimes \cdots \quad \text{and} \quad \det V_0 \otimes \det V_2 \otimes \cdots.$$

Thus Θ determines (uniquely up to sign) a B-isomorphism

$$B \otimes C_\sigma \longrightarrow B$$

which we denote again by Θ.

15.1 Lemma Θ *maps* $1 \otimes C_\sigma \subseteq B \otimes C_\sigma$ *into* $A \subseteq B$.

Proof. We may assume that A is a discrete valuation ring. Let N be the kernel of $\varepsilon : V_0 \to C_\sigma$. Then the first of the exact sequences

$$0 \longrightarrow V_{n+1} \longrightarrow \cdots \longrightarrow V_1 \longrightarrow N \longrightarrow 0 \, , \quad 0 \longrightarrow N \longrightarrow V_0 \overset{\varepsilon}{\longrightarrow} C_\sigma \longrightarrow 0$$

splits over A (N is A-free!). Thus there is already a canonical A-isomorphism

$$\Theta'' \ : \ \det V_1 \otimes \det V_3 \otimes \cdots \quad \longrightarrow \quad \det N \otimes \det V_2 \otimes \cdots$$

and a canonical B-isomorphism

$$\Theta' \ : \ B \otimes C_\sigma \otimes \det N \quad \longrightarrow \quad B \otimes V_0 \, .$$

Then the tensor product $\Theta' \otimes \Theta''$ identifies with $\mathrm{id}_{\det N} \otimes \Theta$ (cf. Supplement 2). Thus it suffices to prove that Θ' maps $1 \otimes C_\sigma \otimes \det N$ into $1 \otimes \det V_0$. Let β be a base for $\det N$ and α a base for $\det V_0$. Then $1 \otimes c \otimes \beta$, $c \in C_\sigma$, is mapped to $w \wedge \beta$, where $w \in B \otimes V_0$ is some element with $(B \otimes \varepsilon)(w) = 1 \otimes c$. Let $v \in V_0$ be any element with $\varepsilon(v) = c$. Then $w - 1 \otimes v \in B \otimes N$, such that $w \wedge \beta = 1 \otimes v \wedge \beta \in A\alpha$. $\qquad\square$

By Lemma 15.1, the composition of $C_\sigma \to 1 \otimes C_\sigma$ with $\Theta : B \otimes C_\sigma \to B$ maps C_σ into A and therefore defines a non-trivial A-linear form

$$\rho = \rho_A^F \ : \ C_\sigma \longrightarrow A \, .$$

Its kernel is the A-torsion of C_σ, that is: $\ker \rho_A^F = \mathfrak{Q}_\sigma$. To compute its image at least in codimension 1 we place ourselves in the situation above where A is a discrete valuation ring. Then we can use the free resolution $0 \to N \to V_0 \overset{\varepsilon}{\longrightarrow} C_\sigma \to 0$. By the calculation at the end of the proof of 15.1

we have $\rho(C_\sigma) \det V_0 = V_0 \wedge \det N \subseteq \det V_0$. Let N' denote the free direct summand $\varepsilon^{-1}(\mathfrak{Q}_\sigma)$ of V_0. Then by definition $\det N = \mathfrak{f}_0(\mathfrak{Q}_\sigma) \det N'$ and $\det V_0 = V_0 \wedge \det N'$, hence $\rho(C_\sigma) \det V_0 = V_0 \wedge \det N = \mathfrak{f}_0(\mathfrak{Q}_\sigma) V_0 \wedge \det N' = \mathfrak{f}_0(\mathfrak{Q}_\sigma) \det V_0$. This proves $\rho(C_\sigma) = \mathfrak{f}_0(\mathfrak{Q}_\sigma)$, which means for an arbitrary integrally closed domain:

$$\operatorname{div} \rho(C_\sigma) \;=\; \operatorname{div} \mathfrak{f}_0(\mathfrak{Q}_\sigma) \;=\; \chi(\mathfrak{Q}_\sigma).$$

Because of 13.1(2) and 13.2 we get the following formula for the divisor of the image $\rho(\Delta)$ of the socle determinant: $\operatorname{div} \rho(\Delta) = \operatorname{div} \rho(A\Delta) = \operatorname{div} \rho(\mathfrak{T}_0 C_\sigma) = \operatorname{div} \mathfrak{T}_0 \rho(C_\sigma) = \operatorname{div} \mathfrak{T}_0 + \operatorname{div} \rho(C_\sigma) = \operatorname{div} \mathfrak{T}_0 + \chi(\mathfrak{Q}_\sigma) = \chi(\bar{C}_\sigma)$. Thus we have proved:

15.2 Theorem *Let A be an integrally closed noetherian domain and let $F_0, \ldots, F_n$ be a regular sequence of homogeneous polynomials in $A[T]$. Then for $C = A[T]/(F_0, \ldots, F_n)$ and $\bar{C}_\sigma = C_\sigma/A\Delta$*

$$\operatorname{div} \rho_A^F(\Delta) \;=\; \chi(\bar{C}_\sigma).$$

In particular, the element $\rho_A^F(\Delta)$ of A generates the resultant ideal $\mathfrak{R}$ and has the same zero set as the elimination ideal $\mathfrak{T}_0$ of C.

The compatibility of the linear form $\rho_A^F : C_\sigma \to A$ with base change which we want to describe now is a special case of a general result, see Supplement 4. In the situation we are discussing here we will use a direct argument.

Let $\varphi : A \to A'$ be a homomorphism of integrally closed noetherian domains such that the images $F_0', \ldots, F_n' \in A'[T]$ of the polynomials $F_0, \ldots, F_n \in A[T]$ form a regular sequence, too (which is the case, for example, if φ is flat).

If $h \in A$ is an element with the same zero set as the elimination ideal $\mathfrak{T}_0 \subseteq A$, the image $h' := \varphi(h) \in A$ has by 10.2 the same zero set as the elimination ideal $\mathfrak{T}_0' \subseteq A'$ corresponding to the polynomials $F_0', \ldots, F_n'$. Therefore $\varphi : A \to A'$ extends to a homomorphism $A_h \to A'_{h'}$ of the rings which are used to construct $\rho = \rho_A^F$ and $\rho' = \rho_{A'}^{F'}$. This gives a commutative diagram

$$
\begin{array}{ccc}
A_h \otimes C_\sigma & \xrightarrow{\;\Theta\;} & A_h \\
\downarrow & & \downarrow \\
A'_{h'} \otimes C'_\sigma & \xrightarrow{\;\Theta'\;} & A'_{h'}
\end{array}
$$

if we order the monomial bases of the modules $V_0, \ldots, V_{n+1}$ and the modules $V_0', \ldots, V_{n+1}'$, $V_i' := A' \otimes V_i$, in the same way. (Otherwise the diagram may be commutative only up to sign.) Restricting Θ and Θ' to $1 \otimes C_\sigma$ and $1 \otimes C_\sigma'$

we get the commutative diagram

$$
\begin{array}{ccc}
C_\sigma & \xrightarrow{\ \rho\ } & A \\
\downarrow & & \downarrow \\
C'_\sigma & \xrightarrow{\ \rho'\ } & A'
\end{array}
$$

such that $\rho' = A' \otimes \rho$. In particular:

15.3 Theorem *Let $\varphi : A \to A'$ be a homomorphism of integrally closed noetherian domains and $\varphi[T] : A[T] \to A'[T]$ its extension. Then for corresponding regular sequences of homogeneous polynomials $F_0, \ldots, F_n \in A[T]$ and $F'_0 = \varphi[T](F_0), \ldots, F'_n = \varphi[T](F_n) \in A'[T]$*

$$
\rho^{F'}_{A'}(\Delta') \ = \ \pm\varphi(\rho^F_A(\Delta))\,.
$$

Theorem 15.3 allows one to define resultants in a very general way. First we observe that any regular sequence $F_0, \ldots, F_n \in A[T]$ over an integrally closed noetherian domain is a specialization of the sequence of generic polynomials

$$
\sum\nolimits_{\langle \nu, \gamma \rangle = \delta_j} U_{j\nu} T^\nu, \quad j = 0, \ldots, n,
$$

in $\mathbb{Z}[U_{j\nu}][T]$, which is also regular by Lemma 8.1(1''). Therefore, by 15.3, *the resultant* $\mathrm{R}(F_0, \ldots, F_n) = \rho^F_A(\Delta)$ *of any such sequence* $F_0, \ldots, F_n$ *is the specialization of the resultant of the corresponding generic polynomials.*

Now let A be an *arbitrary commutative ring* and $F_0, \ldots, F_n$ arbitrary homogeneous polynomials of positive degrees $\delta_0, \ldots, \delta_n$ in the graded polynomial algebra $A[T] = A[T_0, \ldots, T_n]$, where the indeterminates T_i have positive weights $\gamma_0, \ldots, \gamma_n$.

Definition The r e s u l t a n t

$$
R \ = \ \mathrm{R}(F_0, \ldots, F_n) \ = \ \mathrm{R}^\gamma_\delta(F_0, \ldots, F_n) \in A
$$

of the polynomials $F_0, \ldots, F_n$ is the specialization of the resultant of the sequence of the corresponding (saturated) generic polynomials. If this sequence fails to be regular, i.e. if $\gamma = (\gamma_0, \ldots, \gamma_n)$ and $\delta = (\delta_0, \ldots, \delta_n)$ are not admissible, the resultant is zero.

Note that we have not specified the sign of the resultant. As we have seen above, the resultant $R = \mathrm{R}(F_0, \ldots, F_n)$ coincides in case of a regular sequence $F_0, \ldots, F_n$ over an integrally closed noetherian domain A with the element $\pm\rho^F_A(\Delta)$ constructed above. The sign of R can be fixed for a given pair γ, δ by choosing once and for all a monomial base for the construction of ρ. Then equality holds in Theorem 15.3 and one has $\mathrm{R}^\gamma_\delta(F_0, \ldots, F_n) = \rho^F_A(\Delta)$ for all regular sequences $F_0, \ldots, F_n$.

In the classical case $\gamma_0 = \cdots = \gamma_n = 1$ *the resultant R is the classical resultant*, which is defined as the specialization of the generator of the elimination ideal $\mathfrak{T}_0 \subseteq \mathbb{Z}[U_{j\nu}]$ of the ideal generated by the generic polynomials. But in the classical case the elimination ideal $\mathfrak{T}_0$ coincides by 14.4 with the resultant ideal $\mathfrak{R}$ which is by definition generated by the resultant $\rho(\Delta)$. Usually the sign of R is fixed by the rule $R(T_0^{\delta_0}, \ldots, T_n^{\delta_n}) = 1$.

Another point of interest is: If the sequence $F_0, \ldots, F_n \in A[T]$ is the specialization of a regular sequence $F_0'', \ldots, F_n'' \in A''[T]$, where A'' is an integrally closed noetherian domain, $R(F_0, \ldots, F_n) \in A$ is the specialization of the resultant $R(F_0'', \ldots, F_n'') \in A''$, which by Theorem 15.3 can be computed over A'' as the element $\rho_{A''}^{F''}(\Delta'')$ without recurring to the saturated generic case. In particular, one can take for $F_0'', \ldots, F_n''$ any sequence of (not necessarily saturated) generic homogeneous polynomials which can be specialized to $F_0, \ldots, F_n$, as long as this sequence of generic polynomials is admissible. See also Supplement 5 for further results in this direction.

Because in the generic case the zero set of the resultant is the zero set of the elimination ideal, by 10.2 the same holds in general. *Thus for every noetherian ring A the zero set of the resultant $R_\delta^\gamma(F_0, \ldots, F_n)$ is the image in $\mathrm{Spec}\,A$ of the zero set $V_+(F_0, \ldots, F_n) \subseteq \mathbb{P}_\gamma^n(A)$ of the polynomials $F_0, \ldots, F_n$.* (This is also true if A is not noetherian, as is easily seen.) In particular:

15.4 Theorem *Let A be a noetherian ring and let $F_0, \ldots, F_n \in A[T] = A[T_0, \ldots, T_n]$ be homogeneous polynomials of positive degrees. The following conditions are equivalent:*

(1) *The graded A-algebra $A[T]/(F_0, \ldots, F_n)$ is finite, i. e. the extension $A[T]/(F_0, \ldots, F_n)$ of A is a complete intersection of relative dimension 0.*

(2) *The resultant $R(F_0, \ldots, F_n)$ is a unit in A.*

15.5 Corollary *For a regular sequence $F_0, \ldots, F_n$ of homogeneous polynomials of positive degrees in $A[T]$, the resultant $R(F_0, \ldots, F_n)$ is a nonzero-divisor in A.*

Proof. By Proposition 7.4 the residue class ring $Q(A)[T]/(F_0, \ldots, F_n)$ is a complete intersection of relative dimension 0 over the total quotient ring $Q(A)$ of A. By Theorem 15.4 the resultant is a unit in $Q(A)$, i e. a nonzero-divisor in A. $\qquad\qquad\square$

Theorem 15.4 can be generalized to graded complete intersections of arbitrary relative dimension. To do this we introduce the concept of the eliminant (or U-resultant). Let $F_0, \ldots, F_r$, $r \leq n$, be homogeneous polynomials of positive degrees $\delta_0, \ldots, \delta_r$ in the polynomial algebra $A[T] = A[T_0, \ldots, T_n]$ which is graded with respect to the positive weights $\gamma_0, \ldots, \gamma_n$

of the indeterminates $T_0, \ldots, T_n$. Let $\delta := (\delta_0, \ldots, \delta_r)$, $\gamma := (\gamma_0, \ldots, \gamma_n)$, $m := \operatorname{lcm} \gamma$ and

$$G_l := \sum_{i=0}^{n} U_{li} T_i^{m/\gamma_i}, \quad l = 1, \ldots, n - r.$$

Note that all polynomials G_l are homogeneous of degree m in $T_0, \ldots, T_n$. Then the resultant

$$\begin{aligned} \mathrm{R}(F_0, \ldots, F_r) &= \mathrm{R}_\delta^\gamma(F_0, \ldots, F_r) \\ &:= \mathrm{R}_{\delta, m, \ldots, m}^\gamma(F_0, \ldots, F_r, G_1, \ldots, G_{n-r}) \in A[U_{li}]_{l,i} \end{aligned}$$

is called the **eliminant** or U-**resultant** of the sequence $F_0, \ldots, F_r$.

15.6 Corollary *Let A be a noetherian ring and let $F_0, \ldots, F_r$, $r \leq n$, be homogeneous polynomials of positive degrees in $A[T] = A[T_0, \ldots, T_n]$. The following conditions are equivalent:*

(1)   *The graded A-algebra $A[T]/(F_0, \ldots, F_r)$ is a complete intersection of relative dimension $n - r$.*

(2)   *The eliminant $\mathrm{R}(F_0, \ldots, F_r)$ is a primitive polynomial in $A[U_{li}]_{l,i}$.*

Proof. By 7.7, condition (1) on the A-algebra $A[T]/(F_0, \ldots, F_r)$ holds if and only if the $A(U_{li})$-algebra $A(U_{li})[T]/(F_0, \ldots, F_r, G_1, \ldots, G_{n-r})$ is finite. Now Theorem 15.4 can be applied.   □

The proof of 15.6 with the help of 7.7 shows that instead of the polynomials $G_1, \ldots, G_{n-r}$ other generic polynomials could be used to construct eliminants which provide suitable numerical criteria for graded complete intersections, cf. the remark after the proof of 7.7.

Supplements

1. Let B be an arbitrary noetherian ring and

$$0 \longrightarrow W_n \xrightarrow{f_n} \cdots \xrightarrow{f_2} W_1 \xrightarrow{f_1} W_0 \longrightarrow 0$$

an exact sequence of finite projective B-modules. To this sequence there corresponds a canonical isomorphism

$$\Theta : \bigotimes_{i \text{ odd}} \det W_i \xrightarrow{\;\sim\;} \bigotimes_{i \text{ even}} \det W_i$$

of invertible B-modules (which we had to use in this section to construct the mapping $\rho : C_\sigma \to A$). The construction of Θ is compatible with arbitrary changes of base rings.

Let W_{n+1}, W_{-1} be zero modules and $f_{n+1} : W_{n+1} \to W_n$, $f_0 : W_0 \to W_{-1}$ zero mappings. The given sequence splits. Therefore there is an exact sequence of homomorphisms $g_i : W_i \to W_{i+1}$, $i = -1, \ldots, n$, such that

$$f_{i+1} g_i + g_{i-1} f_i = \operatorname{id}_{W_i}$$

for $i = 0, \ldots, n$. We set

$$U := \bigoplus_{i \text{ odd}} W_i, \quad G := \bigoplus_{i \text{ even}} W_i, \quad W := \bigoplus_i W_i = U \oplus G.$$

Then $f := \sum_i f_i$, $g := \sum_i g_i$ are endomorphisms of W such that $f^2 = g^2 = 0$ and
$$fg + gf = \mathrm{id}_W .$$
Now $\Phi := f + g$ is an involution of W with $\Phi(U) = G$ and $\Phi(G) = U$. By the way, $\det \Phi = (-1)^{\mathrm{rk}\, U}$ where $\mathrm{rk}\, U = \mathrm{rk}\, G$ associates to every $\mathfrak{p} \in \mathrm{Spec}\, B$ the rank of $U_\mathfrak{p}$. (For a proof one uses local bases.) The canonical isomorphism Θ we are looking for is
$$\det(\Phi|U) \;:\; \det U = \bigotimes_{i\ \mathrm{odd}} \det W_i \;\longrightarrow\; \bigotimes_{i\ \mathrm{even}} \det W_i = \det G .$$
Indeed, this isomorphism is independent of the chosen sequence g_i, $i \geq 0$. Let $g_i' : W_i \to W_{i+1}$ be another choice of homomorphisms with $f_{i+1}g_i' + g_{i-1}'f_i = \mathrm{id}$, $i = 0,\ldots,n$. For $g' := \sum_i g_i'$, $\Phi' := f + g'$, $h := fg + g'f$, $h' := fg' + gf$ the following equations hold: $hh' = h'h = \mathrm{id}$, $h\Phi = \Phi'h$ and $\det(h|W_i) = 1$ for all i. It follows that $\det(\Phi|U) = \det(h|W) \cdot \det(\Phi|U) = \det((h\Phi)|U) = \det((\Phi'h)|U) = \det(\Phi'|U) \cdot \det(h|U) = \det(\Phi'|U)$. The inverse of $\det(\Phi|U)$ is the canonical isomorphism $\det(\Phi|G)$.

2. Continuing Suppl. 1 and its constructions we consider two exact sequences
$$0 \longrightarrow W_n \xrightarrow{f_n} W_{n-1} \xrightarrow{f_{n-1}} \cdots \xrightarrow{f_{m+1}} W_m \xrightarrow{f''} N \longrightarrow 0,$$
$$0 \longrightarrow N \xrightarrow{f'} W_{m-1} \xrightarrow{f_{m-1}} \cdots \xrightarrow{f_2} W_1 \xrightarrow{f_1} W_0 \longrightarrow 0$$
of finite projective B-modules. They define the exact sequence
$$0 \longrightarrow W_n \longrightarrow \cdots \longrightarrow W_m \xrightarrow{f_m} W_{m-1} \longrightarrow \cdots \longrightarrow W_0 \longrightarrow 0$$
where $f_m := f'f''$. Let two sequences $g'', g_m, \ldots, g_{n-1}$ and $g_0, \ldots, g_{m-1}, g'$ corresponding to the first sequences be fixed in the way described in Suppl. 1. Then $g_0, \ldots, g_{n-1}$ with $g_m := g''g'$ is a sequence corresponding to $f_1, \ldots, f_n$. We consider $N, W_0, \ldots, W_n$ as submodules of a module Z which is the direct sum of these modules. Let X and Y denote the submodules
$$X := \cdots \oplus W_{m-2} \oplus W_m \oplus W_{m+2} \oplus \cdots ,$$
$$Y := \cdots \oplus W_{m-3} \oplus W_{m-1} \oplus W_{m+1} \oplus \cdots$$
of Z. Then the restriction Φ of $f + g = f_1 + \cdots + f_n + g_0 + \cdots + g_{n-1}$ to X is an isomorphism from X to Y, the determinant of which can be computed from the analogous determinants of the isomorphisms Φ' from $X' := \cdots \oplus W_{m-2} \oplus N$ to $Y' := \cdots \oplus W_{m-3} \oplus W_{m-1}$ and Φ'' from $X'' := W_m \oplus W_{m+2} \oplus \cdots$ to $Y'' := N \oplus W_{m+1} \oplus \cdots$. Simply put, $\det \Phi$ can be identified with the product of $\det \Phi'$ and $\det \Phi''$. This follows from the identity
$$\mathrm{id}_N \oplus \Phi =$$
$$(-\mathrm{id}_N \oplus \mathrm{id}_Y) \circ (\mathrm{id}_{N \oplus Y} + f') \circ (\Phi' \oplus \Phi'') \circ (\mathrm{id}_{N \oplus X} - g'') \circ (\mathrm{id}_{N \oplus X} + f'')$$
which is easily checked. The determinants of $\mathrm{id}_{N \oplus Y} + f'$, $\mathrm{id}_{N \oplus X} - g''$, $\mathrm{id}_{N \oplus X} + f''$ are 1. The determinant of $-\mathrm{id}_N \oplus \mathrm{id}_Y$ is $(-1)^{\mathrm{rk}\, N} \in B$. The identification of $\det(\Phi' \oplus \Phi'')$ with $\det \Phi' \cdot \det \Phi''$ requires $(-1)^{\mathrm{rk}\, N}$ as correction factor. (Note that $\sum_{i=0}^{m-1} \mathrm{rk}\, W_i \equiv \mathrm{rk}\, N \bmod 2$.)

3. The construction given in Supplement 1 can be generalized as follows. We consider an exact sequence
$$0 \longrightarrow W_n \xrightarrow{f_n} \cdots \xrightarrow{f_2} W_1 \xrightarrow{f_1} W_0 \xrightarrow{f_0} M \longrightarrow 0$$
over an arbitrary noetherian ring B with finite projective modules $W_0, \ldots, W_n$

and a *torsion module* M. As in Supplement 1 we define $U := \bigoplus_{i \text{ odd}} W_i$ and $G := \bigoplus_{i \text{ even}} W_i$. Over the total quotient ring $Q(B)$ of B the exact sequence

$$0 \longrightarrow Q(B) \otimes W_n \longrightarrow \cdots \longrightarrow Q(B) \otimes W_1 \longrightarrow Q(B) \otimes W_0 \longrightarrow 0$$

yields by Supplement 1 a canonical $Q(B)$-isomorphism

$$\Theta \; : \; Q(B) \otimes (\det U) \xrightarrow{\;\sim\;} Q(B) \otimes (\det G).$$

a) Θ maps the B-submodule $\det U = 1 \otimes (\det U)$ into the B-submodule $\det G = 1 \otimes (\det G)$ and defines therefore an injective B-homomorphism

$$\Theta \; : \; \det U \longrightarrow \det G.$$

(To prove the inclusion $\Theta(\det U) \subseteq \det G$ one reduces to the case that B is a local ring of homological codimension ≤ 1, because B can be identified with the "intersection" of the localizations $B_{\mathfrak{p}}$ of homological codimension ≤ 1. However, in this case $N := \ker f_0 = \operatorname{im} f_1$ is also free because the homological dimension of M is finite and hence ≤ 1. From the exact sequences

$$0 \longrightarrow W_n \longrightarrow \cdots \longrightarrow W_1 \longrightarrow N \longrightarrow 0\,, \quad 0 \longrightarrow N \xrightarrow{\;\iota\;} W_0 \longrightarrow M \longrightarrow 0$$

one gets the result using the method of Suppl. 2, since for the second sequence Θ coincides with $\det \iota$.)

b) The homomorphism $\Theta : \det U \to \det G$ defines a (positive) Cartier divisor D. (By the way, this divisor is independent of the chosen finite projective resolution of M and is simply called the divisor $D = \operatorname{div} M$ of M. See also the next supplement for further information.) If $b \in B$ is an element such that $\det U_b$ and $\det G_b$ are free B_b-modules with base elements u, g, then the divisor D on $\operatorname{Spec} B_b \subseteq \operatorname{Spec} B$ is given by the non-zero-divisor $d \in B_b$ with $\Theta(u) = d \cdot g$.

In particular, if $\det U$ and $\det G$ are globally free with base elements u, g, then D is the principal divisor $\operatorname{div} d$ defined by the equation $\Theta(u) = d \cdot g$. This situation occurs, for example, if B is factorial or if $W_0, \ldots, W_n$ are free.

The invertible ideal $\mathfrak{d}$ $(= \mathfrak{d}(M))$ in B defined by the positive divisor D can be characterized by the equation

$$\Theta(\det U) \; = \; \mathfrak{d} \cdot \det G.$$

In particular, the B-module $\mathfrak{d}$ is isomorphic to $(\det U) \otimes (\det G)^{-1}$ (which is nothing else but $\prod_{i=0}^{n} (\det W_i)^{(-1)^{i+1}}$). In case $\det U$ and $\det G$ are free we have $\mathfrak{d} = Bd$ with the non-zero-divisor $d \in B$ described above.

c) The homomorphism Θ constructed in a) is compatible with changes of base rings, i.e. if $B \to B'$ is a homomorphism of noetherian rings such that the sequence

$$0 \longrightarrow W'_n \longrightarrow \cdots \longrightarrow W'_1 \longrightarrow W'_0 \longrightarrow M' \longrightarrow 0,$$

where $W'_i = B' \otimes W_i$ and $M' = B' \otimes M$, is exact (note that in this case M' is necessarily a torsion module over B'), then the canonical diagram

$$
\begin{array}{ccc}
\det U & \xrightarrow{\;\Theta\;} & \det G \\
\downarrow & & \downarrow \\
\det U' & \xrightarrow{\;\Theta'\;} & \det G'
\end{array}
$$

is commutative.

(Consider $\mathfrak{p}' \in \operatorname{Spec} B'$ and denote by $\mathfrak{p}$ the corresponding element in $\operatorname{Spec} B$.

Then $M_{\mathfrak{p}} = 0$ is equivalent with $M'_{\mathfrak{p}'} = 0$ and for these prime ideals $\mathfrak{p}'$ the diagram

$$\begin{array}{ccc} \det U_{\mathfrak{p}} & \xrightarrow{\;\Theta_{\mathfrak{p}}\;} & \det G_{\mathfrak{p}} \\ \downarrow & & \downarrow \\ \det U'_{\mathfrak{p}'} & \xrightarrow{\;\Theta'_{\mathfrak{p}'}\;} & \det G'_{\mathfrak{p}'} \end{array}$$

is commutative for trivial reasons. This implies the commutativity of the original diagram.)

As a consequence, for the invertible ideals $\mathfrak{d} = \mathfrak{d}(M)$ and $\mathfrak{d}' = \mathfrak{d}(M')$, associated to Θ and Θ' according to b), the following simple equality holds:

$$\mathfrak{d}' = B'\mathfrak{d} .$$

(Remark. The constructions in this supplement can be extended to a whole theory of determinants and divisors for complexes. For this we refer to the article [26] by F. Knudsen and D. Mumford and to the references cited there.)

4. Let A be an integrally closed noetherian domain. Let M be a finite torsion module over A with a finite resolution

$$0 \longrightarrow W_n \longrightarrow \cdots \longrightarrow W_1 \xrightarrow{f_1} W_0 \xrightarrow{f_0} M \longrightarrow 0$$

where all the W_i are finite projective A-modules. Let D denote the positive divisor which belongs to this resolution according to Suppl. 3b) above and $\mathfrak{d}$ the corresponding invertible ideal. Then D, interpreted as a Weil divisor, coincides with the ordinary divisor $\chi(M)$ of M, i.e.

$$D = \chi(M) = \sum_{\mathrm{ht}\,\mathfrak{p}=1} \mathrm{length}_{A_{\mathfrak{p}}}(M_{\mathfrak{p}}) \cdot \mathfrak{p} = \mathrm{div}\,f_0(M)$$

or, on the level of divisorial ideals, $\mathfrak{d} = \mathfrak{div}\,f_0(M)$. (One reduces to the case that A is a discrete valuation ring. Then $N := \ker f_0 = \mathrm{im}\,f_1 \subseteq W_0$ is free.)

Now we assume that $A \to A'$ is a homomorphism of A into another integrally closed noetherian domain A' and that the sequence

$$0 \longrightarrow W'_n \longrightarrow \cdots \longrightarrow W'_1 \longrightarrow W'_0 \longrightarrow M' \longrightarrow 0$$

obtained from the original resolution of M by change of base rings is exact, too. Let D' denote the corresponding divisor and $\mathfrak{d}'$ the corresponding invertible ideal. Then $\mathfrak{d}' = B'\mathfrak{d}$ and hence

$$\mathfrak{div}\,f_0(M') = B'\mathfrak{div}\,f_0(M)$$

(Observe that the elementary equality $f_0(M') = B'f_0(M)$ holds for any change of base rings $A \to A'$; the corresponding equality for the divisorial ideals is by no means obvious.)

5. Let A be an integrally closed noetherian domain, $F_0, \ldots, F_n$ a regular sequence of homogeneous polynomials of positive degrees $\delta_0, \ldots, \delta_n$ in the polynomial algebra $A[T] = A[T_0, \ldots, T_n]$ graded by the weights $\gamma_0, \ldots, \gamma_n$ on the indeterminates and $C := A[T]/(F_0, \ldots, F_n)$, $\bar{C} := C/\mathfrak{T}(C)$. For $r \in \mathbb{N}$ we denote by $\mathfrak{R}_r$ the divisorial ideal

$$\mathfrak{R}_r = \mathfrak{R}_r(F_0, \ldots, F_n) := \mathfrak{d}(\bar{C}_r) = \mathfrak{div}\,f_0(\bar{C}_r) .$$

Note that $\mathfrak{R}_0$ coincides with the elimination ideal $\mathfrak{T}_0$ and is in general not invertible, cf. Sect. 13, Suppl. 1. The ideals $\mathfrak{R}_r$ are principal for $r \geq \sigma = (\delta_0 + \cdots + \delta_n) - (\gamma_0 + \cdots + \gamma_n)$ because the A-modules $\bar{C}_\sigma = C_\sigma/A\Delta$ and $\bar{C}_r = C_r$ for $r > \sigma$ have finite free resolutions by means of the Koszul resolution of C.

By definition, $\mathfrak{R}_\sigma$ is the resultant ideal $\mathfrak{R} = \mathfrak{R}(F_0, \ldots, F_n)$ and is generated by the resultant $R = R(F_0, \ldots, F_n)$. By 14.7, $\mathfrak{R}_r = \mathfrak{R}$ for all $r > \sigma$ with $r \in N_C^u$.

If $\mathfrak{R}_r(F_0, \ldots, F_n) = \mathfrak{R}_s(F_0, \ldots, F_n)$ for some $r, s \geq \sigma$ then $\mathfrak{R}_r(F_0', \ldots, F_n') = \mathfrak{R}_s(F_0', \ldots, F_n')$ for any homomorphism $\varphi : A \to A'$ of integrally closed noetherian domains with the property that the sequence $F_j' := \varphi[T](F_j)$, $j = 0, \ldots, n$, is regular in $A'[T]$. (This follows from Suppl. 4.) In particular: If $\mathfrak{R}_r(F_0, \ldots, F_n) = \mathfrak{R}(F_0, \ldots, F_n)$ for some $r > \sigma$ then $\mathfrak{R}_r(F_0', \ldots, F_n') = \mathfrak{R}(F_0', \ldots, F_n')$.

This result can be used to compute resultants by going back to a generic situation. For example, $\mathfrak{R}_r(F_0, \ldots, F_n) = \mathfrak{R}(F_0, \ldots, F_n)$ for all $r > \sigma$ with $r \in N_S^u$, where S is the generic algebra $Q[T]/(F_0, \ldots, F_n)$ with the generic polynomials $F_j := \sum_{\nu \in N_{\delta_j}(\gamma)} U_{j\nu} T^\nu$, $j = 0, \ldots, n$. Cf. the proof of 14.8.

16 Formulas on Resultants

In this section we develop some characteristic properties of resultants which are often used in calculations.

To begin with, we determine their degrees in the generic case. Let $\gamma_0, \ldots, \gamma_n$ be the positive weights of the indeterminates $T_0, \ldots, T_n$ and let

$$F_j = \sum_{\nu \in N_j} U_{j\nu} T^\nu \in Q[T],$$

$N_j \subseteq N_{\delta_j}(\gamma)$, $Q := \mathbb{Z}[U_{j\nu}]_{j,\nu}$, be an admissible sequence of generic homogeneous polynomials of degrees $\delta_j > 0$, $j = 0, \ldots, n$. We consider the Q-algebra $C = Q[T]/(F_0, \ldots, F_n)$ and its Poincaré series

$$\mathcal{P}_C = \sum_{m \geq 0} c_m Z^m = \frac{(1 - Z^{\delta_0}) \cdots (1 - Z^{\delta_n})}{(1 - Z^{\gamma_0}) \cdots (1 - Z^{\gamma_n})},$$

$c_m := \operatorname{rank}_Q C_m$, which is a polynomial of degree $\sigma = \sum_j \delta_j - \sum_i \gamma_i$. The Poincaré series of $Q[T]$ itself is

$$\mathcal{P}_{Q[T]} = \sum_{m \geq 0} p_m Z^m = \frac{1}{(1 - Z^{\gamma_0}) \cdots (1 - Z^{\gamma_n})},$$

$p_m := \operatorname{rank}_Q Q[T]_m$. Furthermore, we are going to use the algebras

$$C_{\langle j \rangle} := Q[T]/(F_0, \ldots, F_{j-1}, F_{j+1}, \ldots, F_n),$$

$j = 0, \ldots, n$, and their Poincaré series which can be written in the form

$$\mathcal{P}_j = \mathcal{P}_C/(1 - Z^{\delta_j}).$$

16.1 Degree formula *The generic resultant $R = \mathrm{R}(F_0, \ldots, F_n) \in Q = \mathbb{Z}[U_{j\nu}]_{j,\nu}$ is, for every j, a homogeneous polynomial of positive degree in the indeterminates $U_{j\nu}$, $\nu \in N_j$, which are assumed to have weight 1.*

The degree ε_j in the $U_{j\nu}$, $\nu \in N_j$, can be computed in the following ways:

(1) $$\varepsilon_j = 1 + \sum_{J, \, j \in J} (-1)^{|J|-1} p_{\sigma - \delta_J}.$$

Here, $\delta_J := \sum_{l \in J} \delta_l$ for a subset $J \subseteq \{0, \ldots, n\}$.

(2) $$\varepsilon_j = \sum_{J, \, j \notin J} (-1)^{|J|} p_{\sigma - \delta_J} = \operatorname{rank}_Q (C_{\langle j \rangle})_\sigma.$$

(3) $$\varepsilon_j = \sum_{k \geq 0} c_{k\delta_j}.$$

The total degree of R as a polynomial in all the $U_{j\nu}$ is $\varepsilon_0 + \cdots + \varepsilon_n$.

Proof. We may assume $j = 0$. We adjoin a new indeterminate S to Q and have to show $\tilde{R} := \mathrm{R}(SF_0, F_1, \ldots, F_n) = S^{\varepsilon_0} R$. This can be done in the quotient field L of $Q[S]$.

The Koszul complexes $\mathrm{K}(SF_0, F_1, \ldots, F_n)$ and $\mathrm{K}(F_0, F_1, \ldots, F_n)$ over $L[T]$ are free with $L[T]$-bases $e_J = e_{j_1} \wedge \cdots \wedge e_{j_r}$, $J \subseteq \{0, \ldots, n\}$, $J = \{j_1, \ldots, j_r\}$, $j_1 < \cdots < j_r$, $\deg e_J = \delta_J$. The canonical homomorphism

$$\mathrm{K}(SF_0, F_1, \ldots, F_n) \longrightarrow \mathrm{K}(F_0, F_1, \ldots, F_n)$$

of complexes is defined by $e_J \mapsto Se_J$, if $0 \in J$, and $e_J \mapsto e_J$, if $0 \notin J$. It yields a commutative diagram

$$
\begin{array}{ccccccccc}
0 & \longrightarrow & V_{n+1} & \longrightarrow & \cdots & \longrightarrow & V_0 & \longrightarrow & D_\sigma & \longrightarrow & 0 \\
& & \downarrow & & & & \downarrow & & \downarrow & & \\
0 & \longrightarrow & V_{n+1} & \longrightarrow & \cdots & \longrightarrow & V_0 & \longrightarrow & D_\sigma & \longrightarrow & 0
\end{array}
$$

of L-vector spaces in degree σ with

$$D := L[T]/(F_0, \ldots, F_n) = L[T]/(SF_0, F_1, \ldots, F_n)$$

and therefore a canonical commutative diagram

$$
\begin{array}{ccc}
D_\sigma \otimes \det V_1 \otimes \det V_3 \otimes \cdots & \xrightarrow{\ \tilde{\Theta}\ } & \det V_0 \otimes \det V_2 \otimes \cdots \\
\ \ \downarrow \varphi_u & & \ \ \downarrow \varphi_g \\
D_\sigma \otimes \det V_1 \otimes \det V_3 \otimes \cdots & \xrightarrow{\ \Theta\ } & \det V_0 \otimes \det V_2 \otimes \cdots
\end{array}
$$

of L-vector spaces. Let ω_u be a base of $\det V_1 \otimes \det V_3 \otimes \cdots$ and ω_g a base of $\det V_0 \otimes \det V_2 \otimes \cdots$. Let ρ be the canonical mapping $\rho_L^{F_0, \ldots, F_n} : D_\sigma \to L$ and $\tilde{\rho}$ the canonical mapping $\rho_L^{SF_0, \ldots, F_n} : D_\sigma \to L$. Then

$$\tilde{\Theta}(x \otimes \omega_u) = \tilde{\rho}(x)\omega_g, \quad \Theta(x \otimes \omega_u) = \rho(x)\omega_g,$$

for any $x \in D_\sigma$. Because of $V_r = \bigoplus_{|J|=r} L[T]_{\sigma - \delta_J} e_J$ the mapping φ_u is simply multiplication by S^{ε_u} and similarly, φ_g is multiplication by S^{ε_g},

$$\varepsilon_u := \sum_{0 \in J, \, |J| \text{ odd}} p_{\sigma - \delta_J}, \quad \varepsilon_g := \sum_{0 \in J, \, |J| \text{ even}} p_{\sigma - \delta_J}.$$

Let $\Delta := \Delta_{F_0, \ldots, F_n}^{T_0, \ldots, T_n} \in D_\sigma$ and $\tilde{\Delta} := \Delta_{SF_0, \ldots, F_n}^{T_0, \ldots, T_n} \in D_\sigma$. Obviously, $\tilde{\Delta} = S\Delta$. By definition, $\rho(\Delta) = R$ and $\tilde{\rho}(\tilde{\Delta}) = \tilde{R}$. Now,

$$\varphi_g \circ \tilde{\Theta}(\tilde{\Delta} \otimes \omega_u) = \varphi_g(\tilde{\rho}(\tilde{\Delta})\omega_g) = \varphi_g(\tilde{R}\omega_g) = S^{\varepsilon_g}\tilde{R}\omega_g,$$

$$\Theta \circ \varphi_u(\tilde{\Delta} \otimes \omega_u) = \rho(S^{\varepsilon_u}\tilde{\Delta})\omega_g = S^{\varepsilon_u+1}\rho(\Delta)\omega_g = S^{\varepsilon_u+1}R\omega_g.$$

Thus $\tilde{R} = S^{1+\varepsilon_u-\varepsilon_g}R$. Therefore, R is homogeneous in the indeterminates $U_{0\nu}$ of degree $\varepsilon_0 = 1 + \varepsilon_u - \varepsilon_g$, that is

$$\varepsilon_0 = 1 + \sum_{J, \, 0 \in J} (-1)^{|J|-1} p_{\sigma - \delta_J} = 1 + \sum_{J, \, 0 \notin J} (-1)^{|J|} p_{\sigma - \delta_0 - \delta_J},$$

and $\varepsilon_0 - 1$ is the coefficient of Z^σ in the series

$$\frac{Z^{\delta_0} \cdot (1 - Z^{\delta_1}) \cdots (1 - Z^{\delta_n})}{(1 - Z^{\gamma_0}) \cdots (1 - Z^{\gamma_n})} = Z^{\delta_0} \mathcal{P}_0 .$$

Because the highest coefficient c_σ of $\mathcal{P}$ is one, ε_0 is the coefficient of Z^σ in the series $\mathcal{P} + Z^{\delta_0} \mathcal{P}_0 = \mathcal{P}_0$. This proves equation (2). It remains to prove equation (3) of the theorem. But $\mathcal{P}_0 = \mathcal{P}/(1 - Z^{\delta_0})$ and therefore

$$\varepsilon_0 = \sum_{k \geq 0} c_{\sigma - k\delta_0} = \sum_{k \geq 0} c_{k\delta_0}$$

because of $c_{\sigma - m} = c_m$ for all $m \in \mathbb{N}$, see Supplement 2 of Section 7. $\quad\square$

16.2 Corollary *Assume that the n generic polynomials $F_0, \ldots, \hat{F}_j, \ldots, F_n$ form a strictly admissible sequence for a particular j. Then the resultant $\mathrm{R}(F_0, \ldots, F_n) \in \mathbb{Z}[U_{j\nu}]_{j,\nu}$ is a homogeneous polynomial in the indeterminates $U_{j\nu}$, $\nu \in N_j$, of degree*

$$\varepsilon_j = \gcd(\gamma_0, \ldots, \gamma_n) \cdot \frac{\delta_0 \cdots \hat{\delta}_j \cdots \delta_n}{\gamma_0 \cdots \gamma_n} = \frac{\gcd(\gamma)}{\delta_j} \cdot \frac{\prod_i \delta_i}{\prod_i \gamma_i} .$$

Proof. Let $d := \gcd(\gamma)$. Replacing γ_i by γ_i/d and δ_j by δ_j/d we may assume $d = 1$. Then the Poincaré series $\mathcal{P}_j$ of $C_{(j)}$ gets stationary at $\sigma - \delta_j + 1$ with limit $\prod_{i \neq j} \delta_i / \prod_i \gamma_i$, cf. Supplement 7 of Section 11. $\quad\square$

Corollary 16.2 implies that *in the classical case $\gamma_0 = \cdots = \gamma_n = 1$ the degree of the resultant $\mathrm{R}(F_0, \ldots, F_n)$ with respect to the coefficients of F_j is $\delta_0 \cdots \hat{\delta}_j \cdots \delta_n$*.

We note that the degree ε_j in Theorem 16.1 coincides with the rank of the (finite) Q-algebra $\Gamma(\mathcal{O}_{\mathrm{Proj}\, C_{(j)}})$. This is a special case of the following more general result.

16.3 Theorem *Let A be a noetherian ring and let $F_1, \ldots, F_n \in A[T] = A[T_0, \ldots, T_n]$ be a regular sequence of homogeneous polynomials of positive degrees $\delta_1, \ldots, \delta_n$. Let $D := A[T]/(F_1, \ldots, F_n)$. Then $B := \Gamma(\mathcal{O}_{\mathrm{Proj}\, D})$ is a finite A-algebra of rank*

$$\varepsilon = 1 + \sum_{J \subseteq \{1, \ldots, n\}} (-1)^{|J|} p_{\tau - \delta_J}$$

with $\tau := (\delta_1 + \cdots + \delta_n) - (\gamma_0 + \cdots + \gamma_n)$, $\delta_J := \sum_{l \in J} \delta_l$ and $p_m := \mathrm{rank}_A A[T]_m$, $m \geq 0$.

If D is a complete intersection, then B is a projective A-algebra.

If the pair $\gamma = (\gamma_0, \ldots, \gamma_n)$, $\delta = (\delta_1, \ldots, \delta_n)$ is strictly admissible, then

$$\varepsilon = \gcd(\gamma_0, \ldots, \gamma_n) \cdot \frac{\delta_1 \cdots \delta_n}{\gamma_0 \cdots \gamma_n} .$$

Proof. To prove the first statement about the rank of B we may extend A to its total ring of fractions $Q(A)$. Then, by 7.4, $Q(A) \otimes_A D$ is a complete intersection of relative dimension 1 over $Q(A)$ by 7.4. Furthermore there exists a canonical isomorphism

$$Q(A) \otimes_A B = \Gamma(\mathcal{O}_{\mathrm{Proj}\,Q(A) \otimes_A D}).$$

Thus to prove all parts of the lemma it suffices to consider the special situation that D is a complete intersection over A of relative dimension 1.

Using a Kronecker extension of A we can assume in addition that there is a homogeneous polynomial $G \in A[T]$, the residue class of which in D we denote by g, such that $D/gD = A[T]/(G, F_1, \ldots, F_n)$ is a finite complete intersection, cf. the proof of 15.6. Now simply $B = (D_g)_0$. This proves already that B is flat and hence projective over A.

Let δ_0 (> 0) be the degree of g. Then we have $B = \bigcup_{k \geq 0} D_{k\delta_0}/g^k$ and $D_0 \subseteq D_{\delta_0}/g \subseteq D_{2\delta_0}/g^2 \subseteq \cdots$. Because of

$$\frac{D_{k\delta_0}}{g^k} \bigg/ \frac{D_{(k-1)\delta_0}}{g^{k-1}} = D_{k\delta_0}/gD_{(k-1)\delta_0}$$

the A-rank ε of B is the sum $\sum_{k \geq 0} c_{k\delta_0} = \sum_{k \geq 0} c_{\sigma - k\delta_0}$, where

$$\mathcal{P}_{D/gD} = \frac{(1 - Z^{\delta_0})(1 - Z^{\delta_1}) \cdots (1 - Z^{\delta_n})}{(1 - Z^{\gamma_0})(1 - Z^{\gamma_1}) \cdots (1 - Z^{\gamma_n})} = \sum_{m \geq 0} c_m Z^m$$

is the Poincaré series of D/gD and $\sigma := (\delta_0 + \cdots + \delta_n) - (\gamma_0 + \cdots + \gamma_n)$. This shows that ε is the coefficient of Z^σ in the Poincaré series

$$\mathcal{P}_D = \frac{(1 - Z^{\delta_1}) \cdots (1 - Z^{\delta_n})}{(1 - Z^{\gamma_0}) \cdots (1 - Z^{\gamma_n})} = \frac{\mathcal{P}_{D/gD}}{1 - Z^{\delta_0}} = Z^{\delta_0}\mathcal{P}_D + \mathcal{P}_{D/gD},$$

i.e. the rank of D_σ. Since the coefficient of Z^σ in $\mathcal{P}_{D/gD}$ is 1, the formula for ε follows (as at the end of the proof of 16.1). The last statement about ε follows in the same way as in the proof of Corollary 16.2. $\square$

In the proof of Theorem 16.3 we have seen: If D contains a homogeneous non-zero-divisor g of degree $\delta_0 > 0$, then the A-rank of B is the A-rank of D_σ, $\sigma := (\delta_0 + \cdots + \delta_n) - (\gamma_0 + \cdots + \gamma_n)$. In the case that D/gD is finite, i.e. a finite complete intersection over A, there is more to be said about D_σ with respect to duality theory.

16.4 Lemma *Let A be a noetherian ring and let $F_0, \ldots, F_n \in A[T] = A[T_0, \ldots, T_n]$ be a regular sequence of homogeneous polynomials of positive degrees $\delta_0, \ldots, \delta_n$ such that $C = A[T]/(F_0, \ldots, F_n)$ is a finite complete intersection over A. Let $D = A[T]/(F_1, \ldots, F_n)$, $B = \Gamma(\mathcal{O}_{\mathrm{Proj}\,D})$ and $\sigma = (\delta_0 + \cdots + \delta_n) - (\gamma_0 + \cdots + \gamma_n)$. Then D_σ is a B-module and there is a natural B-isomorphism $D_\sigma \cong \mathrm{Hom}_A(B, A)$.*

Proof. Let f_0 be the residue class of F_0 in D. Since $C_m = 0$ for $m > \sigma$ we have $D_m = f_0 D_{m-\delta_0}$ for all $m > \sigma$ and $B = D_{k\delta_0}/f_0^k$, where k is the largest integer with $k\delta_0 \leq \sigma$. Then $B \cdot D_\sigma = D_{k\delta_0} D_\sigma/f_0^k \subseteq D_{k\delta_0+\sigma}/f_0^k = D_\sigma$, hence D_σ is a B-module.

To prove that D_σ is isomorphic to $\operatorname{Hom}_A(B, A)$ (which is the relative dualizing module of B over A) the A-bilinear map $\langle -, - \rangle : B \times D_\sigma \to A$ defined by $\langle b, d \rangle := \eta(\overline{bd})$, $b \in B$, $d \in D_\sigma$, is used, where $\eta = \eta_{F_0,\ldots,F_n}^{T_0,\ldots,T_n}$ is the canonical A-linear form $C \to A$ constructed at the beginning of Section 12. By definition, $\langle -, - \rangle$ satisfies the equality $\langle b'b, d \rangle = \langle b, b'd \rangle$ for $b', b \in B$, $d \in D_\sigma$. Furthermore, $\langle -, - \rangle$ defines a perfect duality over A and therefore yields the desired isomorphism $D_\sigma \cong \operatorname{Hom}_A(B, A)$.

To prove that $\langle -, - \rangle$ indeed defines a perfect duality we may reduce modulo the maximal ideals of A and may therefore assume that $A = K$ is a field. Because $\dim_K B = \dim_K D_\sigma$ by 16.3, it suffices to show: If $\langle b, D_\sigma \rangle = 0$ for some $b \in B$, i.e. $\overline{bD_\sigma} = 0$ in C or $bD_\sigma \subseteq f_0 D_{\sigma-\delta_0}$, then $b = 0$. Let $b = g/f_0^l$ with minimal $l \leq k$. The assumption on b implies $gD_\sigma \subseteq f_0^{l+1} D_{\sigma-\delta_0}$ and in particular $gf_0^l D_{\sigma-l\delta_0} \subseteq f_0^{l+1} D_{\sigma-\delta_0}$, from which $gD_{\sigma-l\delta_0} \subseteq f_0 D_{\sigma-\delta_0}$ follows. Since the multiplication $C_{l\delta_0} \times C_{\sigma-l\delta_0} \to C_\sigma = A\Delta$ defines a perfect duality over A (see again Section 12), the element g belongs to $f_0 D$, which implies $g = 0$ by the minimality of l. $\qquad\square$

The isomorphism $D_\sigma \cong \operatorname{Hom}_A(B, A)$ constructed in the proof of Lemma 16.4 identifies the A-submodule $D_{\sigma-\delta_0} \cong f_0 D_{\sigma-\delta_0} \subseteq D_\sigma$ with the **reduced dualizing module**

$$\operatorname{Hom}_A(B/A, A) \cong \{\alpha \in \operatorname{Hom}_A(B, A) : \alpha(1) = 0\} \subseteq \operatorname{Hom}_A(B, A).$$

Note that $\tau := \sigma - \delta_0 = (\delta_1 + \cdots + \delta_n) - (\gamma_0 + \cdots + \gamma_n)$ depends only on $F_1, \ldots, F_n$.

The methods of duality used in the proof of 16.4 stem from [42], where the classical case $\gamma_0 = \cdots = \gamma_n = 1$ is treated more extensively in this respect.

The next result is the **norm formula** for resultants, sometimes also called **Poisson formula**.

16.5 Norm formula *Let A be a noetherian ring and let $F_0', F_0'', F_1, \ldots, F_n$ be homogeneous polynomials of positive degrees $\delta_0, \delta_0, \delta_1, \ldots, \delta_n$ in $A[T] = A[T_0, \ldots, T_n]$. Assume that the sequence $F_0'', F_1, \ldots, F_n$ is regular. Let*

$$D = A[T]/(F_1, \ldots, F_n), \quad B = \Gamma(\mathcal{O}_{\operatorname{Proj} D})$$

and let f_0', f_0'' denote the residue classes of F_0', F_0'' in D. Then

$$\frac{\operatorname{R}(F_0', F_1, \ldots, F_n)}{\operatorname{R}(F_0'', F_1, \ldots, F_n)} = \operatorname{N}_A^B\!\left(\frac{f_0'}{f_0''}\right).$$

Proof. Note that by Theorem 16.3 the A-algebra B is finite with a well-defined rank. The element f_0'/f_0'' belongs to the total quotient ring $Q(B) = Q(A) \otimes_A B$ of B, and the norm N_A^B is by definition the norm $N_{Q(A)}^{Q(B)}$ of the (free) $Q(A)$-algebra $Q(B)$.

It goes without saying that the rules for determining the signs of both resultants occurring in the formula have to be the same.

The formula can be verified over the total quotient ring $Q(A)$ of A. Therefore we may assume that $A[T]/(F_0'', F_1, \ldots, F_n)$ is a finite complete intersection. In this case the formula is a specialization of the corresponding formula for generic polynomials. (Note that in this case also B is obtained from the generic situation by change of base rings.)

We now consider the generic situation. Besides the generic polynomials
$$F_0' = \sum_{\nu \in N_0} U_{0\nu}' T^\nu, \quad F_0'' = \sum_{\nu \in N_0} U_{0\nu}'' T^\nu, \quad N_0 = N_{\delta_0}(\gamma),$$
and $F_j = \sum_{\nu \in N_j} U_{j\nu} T^\nu$, $j = 1, \ldots, n$, we are going to use another generic polynomial
$$F_0 = \sum_{\nu \in N_0} U_{0\nu} T^\nu.$$
Then our base ring is
$$Q = \mathbb{Z}[U_{0\nu}', U_{0\nu}'', U_{0\nu}, \nu \in N_0; \; U_{j\nu}, j = 1, \ldots, n, \nu \in N_j].$$
$F_0', F_1, \ldots, F_n$ and $F_0'', F_1, \ldots, F_n$ and $F_0, F_1, \ldots, F_n$ are regular sequences.

It suffices to show that $R(F_0', F_1, \ldots, F_n)/R(F_0'', F_1, \ldots, F_n)$ and the element $N_Q^B(f_0'/f_0'')$ have the same divisor. Then both elements differ just by a factor ± 1, which has to be 1. This can be verified by specializing F_0' and F_0'' to one and the same element, say F_0.

Each of the prime divisors of $R(F_0', F_1, \ldots, F_n)/R(F_0'', F_1, \ldots, F_n)$ and $N_Q^B(f_0'/f_0'')$ contains at least one of the indeterminates $U_{0\nu}'$ or $U_{0\nu}''$, because the automorphism of Q which interchanges $U_{0\nu}' \leftrightarrow U_{0\nu}''$ and leaves the indeterminates $U_{j\nu}$, $j \geq 0$, fixed, transforms both functions into their inverses. Therefore we can establish the equality of their divisors in the ring of fractions $A := Q_{R(F_0, F_1, \ldots, F_n)}$. This reduces the proof to the proof of the equality of the divisors of $R(F_0', F_1, \ldots, F_n)/R(F_0, F_1, \ldots, F_n)$ and $N_Q^B(f_0'/f_0)$ over A, where f_0 is the residue class of F_0 in $D = Q[T]/(F_1, \ldots, F_n)$.

Thus the norm formula is proved in general if we prove the equality
$$\operatorname{div} R(F_0', F_1, \ldots, F_n) \left(= \operatorname{div} \frac{R(F_0', F_1, \ldots, F_n)}{R(F_0'', F_1, \ldots, F_n)} \right) = \operatorname{div} N_A^B \left(\frac{f_0'}{f_0''} \right)$$
in case that $F_0', F_1, \ldots, F_n$ and $F_0'', F_1, \ldots, F_n$ are regular sequences and that A is integrally closed and $C'' := A[T]/(F_0'', F_1, \ldots, F_n)$ is even a finite complete intersection over A.

Let Δ' and Δ'' be the socle determinants corresponding to $F_0', F_1, \ldots, F_n$ and $F_0'', F_1, \ldots, F_n$, respectively. The socle determinant of $F_0' F_0'', F_1, \ldots, F_n$

as an element of $D/f_0' f_0'' D$ is the residue class of $f_0'' \Delta'$ as well as of $f_0' \Delta''$. Therefore, in D we have

$$f_0' f_0'' D_{\sigma - \delta_0} + A f_0'' \Delta' \; = \; f_0' f_0'' D_{\sigma - \delta_0} + A f_0' \Delta''$$

and hence

$$f_0' D_{\sigma - \delta_0} + A \Delta' \; = \; \frac{f_0'}{f_0''} (f_0'' D_{\sigma - \delta_0} + A \Delta'') \; = \; \frac{f_0'}{f_0''} D_\sigma \, ,$$

because of $D_\sigma = f_0'' D_{\sigma - \delta_0} + A \Delta''$ since C'' is a complete intersection (cf. 12.5, for instance). However, the divisor of $\mathrm{R}(F_0', F_1, \ldots, F_n)$ is by definition the divisor d of the A-module $D_\sigma / (f_0' D_{\sigma - \delta_0} + A \Delta')$. Since $f_0' D_{\sigma - \delta_0} + A \Delta' = (f_0' / f_0'') D_\sigma$, the divisor d is the same as the divisor of the A-determinant of the multiplication on D_σ by f_0' / f_0''. Lemma 16.4 now shows that $d = \operatorname{div} \mathrm{N}_A^B (f_0' / f_0'')$. □

The degree formula 16.1 is a consequence of the norm formula 16.5 and of Theorem 16.3, namely: For an arbitrary element $a \in A$ and a regular sequence $F_0, \ldots, F_n$ in $A[T_0, \ldots, T_n]$ one has

$$\mathrm{R}(aF_0, F_1, \ldots, F_n) \; = \; \mathrm{N}_A^B(a) \mathrm{R}(F_0, F_1, \ldots, F_n) \; = \; a^{\varepsilon_0} \mathrm{R}(F_0, F_1, \ldots, F_n)$$

where $\varepsilon_0 = \mathrm{rk}_A B$.

A product formula for resultants does not hold in general, cf. Supplement 2, but can be proved in special situations as in the following theorem.

16.6 Product formula *Let* $F_0', F_0'', F_1, \ldots, F_n$ *be homogeneous polynomials of positive degrees* $\delta_0', \delta_0'', \delta_1, \ldots, \delta_n$ *in* $A[T] = A[T_0, \ldots, T_n]$, *$A$ a noetherian ring. Assume that the pair* $\gamma = (\gamma_0, \ldots, \gamma_n)$, $\delta'' = (\delta_0'', \delta_1, \ldots, \delta_n)$ *is strictly admissible. Then*

$$\mathrm{R}(F_0' F_0'', F_1, \ldots, F_n) \; = \; \pm \mathrm{R}(F_0', F_1, \ldots, F_n) \, \mathrm{R}(F_0'', F_1, \ldots, F_n) \, .$$

Proof. If the pair γ, $\delta' = (\delta_0', \delta_1, \ldots, \delta_n)$ is not admissible, then the pair γ, $\delta = (\delta_0' + \delta_0'', \delta_1, \ldots, \delta_n)$ is not admissible and both sides of the formula are zero. Thus we may assume that γ, δ' is admissible and also that the polynomials $F_0', F_0'', F_1, \ldots, F_n$ are generic polynomials in $Q[T]$. Then we have to prove that the divisors of the elements on both sides of the product formula coincide, i. e.

$$\operatorname{div} \mathrm{R}(F_0' F_0'', F_1, \ldots, F_n) = \operatorname{div} \mathrm{R}(F_0', F_1, \ldots, F_n) + \operatorname{div} \mathrm{R}(F_0'', F_1, \ldots, F_n) \, .$$

Let $D := Q[T]/(F_1, \ldots, F_n)$ and $\tau := (\delta_1 + \cdots + \delta_n) - (\gamma_0 + \cdots + \gamma_n)$. We denote by f_0', f_0'' the residue classes of F_0', F_0'' in D and by Δ' the socle determinant of $F_0', F_1, \ldots, F_n$. Then $f_0'' \Delta'$ is a socle determinant of $F_0' F_0'', F_1, \ldots, F_n$. The divisor of $\mathrm{R}(F_0' F_0'', F_1, \ldots, F_n)$ is the divisor of the

Q-module $D_{\tau+\delta_0'+\delta_0''}/(f_0'f_0''D_\tau + Qf_0''\Delta')$ and hence the sum of the divisors of the modules $D_{\tau+\delta_0'+\delta_0''}/f_0''D_{\tau+\delta_0'}$ and $f_0''D_{\tau+\delta_0'}/f_0''(f_0'D_\tau + Q\Delta') \cong D_{\tau+\delta_0'}/(f_0'D_\tau + Q\Delta')$. The second summand is by definition the divisor of $R(F_0', F_1, \ldots, F_n)$. The first summand coincides by 14.8 with the divisor of $R(F_0'', F_1, \ldots, F_n)$.   □

When the sign of the resultants occurring in 16.6 are fixed for the pairs (γ, δ'), (γ, δ'') and (γ, δ), then the sign in the product formula is well-defined.

Supplements

1. The result of Theorem 16.1 has a straightforward generalization which can be formulated as follows: Let $F_0, \ldots, F_n$ be homogeneous polynomials of positive degrees $\delta_0, \ldots, \delta_n$ in a graded polynomial algebra $A[T] = A[T_0, \ldots, T_n]$, where the weights of the indeterminates T_i are γ_i, $i = 0, \ldots, n$, and where γ, δ are admissible. Assume that the degrees δ_j coincide for a subset $J \subseteq \{0, \ldots, n\}$ and denote by d their common value, by $\varepsilon = \sum_{k \geq 0} c_{kd}$ the common partial degree of the resultant, as described in 16.1. Let

$$\tilde{F}_j = \sum_{i \in J} s_{ij} F_i, \quad s_{ij} \in A, \quad j \in J,$$

and $\tilde{F}_j = F_j$ for $j \notin J$. Then

$$R(\tilde{F}_0, \ldots, \tilde{F}_n) = (\det s_{ij})^\varepsilon R(F_0, \ldots, F_n).$$

(One considers the generic case, adds indeterminates S_{ij}, $(i, j) \in J \times J$, and follows the lines of the proof of 16.1.)

2. Let $A = \mathbb{Z}[U]$ and $\gamma = (1, 2)$. Then $R(T_0, T_0^2 + UT_1) = U = R(T_0^2, T_0^2 + UT_1)$. Thus the product formula does not hold in this simple case.

3. (*Reduction formula*) Let $F_0, F_0', F_1, \ldots, F_n$ be homogeneous polynomials of positive degrees $\delta_0, \delta_0', \delta_1, \ldots, \delta_n$ in $A[T] = A[T_0, \ldots, T_n]$, where $\delta_0 = \delta_0'$. If $F_0 \equiv F_0' \mod (F_1, \ldots, F_n)$ then $R(F_0, F_1, \ldots, F_n) = R(F_0', F_1, \ldots, F_n)$.

References

1. Atiyah, M.F. and Macdonald, I.G.: Introduction to commutative algebra. Addison-Wesley, London 1969

2. Barucci, V., Dobbs, D.E. and Fontana, M.: Maximal properties in numerical semigroups and applications to one-dimensional analytically irreducible local domains. Mem. Amer. Math. Soc. **126**, Nr. 598 (1997)

3. Böger, E. and Storch, U.: A remark on Bezout's theorem. Univ. Iag. acta math. **37**, 25-36 (1999)

4. Bourbaki, N.: Commutative algebra. Hermann, Paris 1972

5. Brauer, A. and Shockley, J.E.: On a problem of Frobenius. J. reine u. angew. Math. **211**, 215-220 (1962)

6. Bruns, W. and Herzog, J.: Cohen-Macaulay rings. Cambr. stud. in adv. math. **39**. Cambridge Univ. Press ²1998

7. Campillo, A. and Pisón, P.: L'idéal d'un semi-groupe de type fini. C. R. Acad. Sci. Paris **316**, Sér. I, 1303-1306 (1993)

8. Chardin, M.: The resultant via a Koszul complex. In: Computational algebraic geometry. Eds. F. Eyssette, A. Galligo. Birkhäuser, Basel, 29-39 (1993)

9. Cox, D., Little, J. and O'Shea, D.: Using algebraic geometry. Grad. Texts in Math. **185**. Springer, New York 1998

10. Davison, J.L.: On the linear Diophantine problem of Frobenius. J. Number Th. **48**, 353-363 (1994)

11. Delorme, C.: Sous-monoïdes d'intersection complète de **N**. Ann. scient. Éc. Norm. Sup., 4ᵉ série, **9**, 145-154 (1976)

12. Eisenbud, D.: Commutative algebra, with a view toward algebraic geometry. Grad. Texts in Math. **150**. Springer, New York 1995

13. Fröberg, R., Gottlieb, C. and Häggkvist, R.: Semigroups, semigroup rings and analytically irreducible rings. Reports Dept. Math. Univ. Stockholm, No. 1 (1986)

14. Gel'fand, I.M., Kapranov, M.M. and Zelevinsky, A.V.: Discriminants, resultants and multidimensional determinants. Mathematics: Theory and Applications. Birkhäuser, Boston 1994

15. Hartshorne, R.: Algebraic geometry. Grad. Texts in Math. **52**. Springer, New York 1977

16. Herzog, J.: Generators and relations of abelian semigroups and semigroup rings. Manuscr. math. **3**, 175-193 (1970)

17. Herzog, J. and Kunz, E.: Die Wertehalbgruppe eines lokalen Rings der Dimension 1. Sitzungsber. Heidelberger Akad. d. Wiss. 27-67 (1970)

18. Hilton, P.J. and Stammbach, U.: A course in homological algebra. Grad. Texts in Math. **4**. Springer, New York ²1997

19. Hurwitz, A.: Über die Trägheitsformen eines algebraischen Moduls. Annali Mat. pur. appl., serie III, **20**, 113-151 (1913). Cf.: Math. Werke von Adolf Hurwitz, II, 591-626. Birkhäuser, Basel 1933

20. Johnson, S.M.: A linear Diophantine problem. Can. J. Math. **12**, 390-398 (1960)

21. Jouanolou, J.P.: Idéaux résultants. Adv. Math. **37**, 212-238 (1980)

22. —: Le formalisme du résultant. Adv. Math. **90**, 117-263 (1991)

23. —: Aspects invariant de l'élimination. Adv. Math. **114**, 1-174 (1995)

24. —: Formes d'inertie et résultant: un formulaire. Adv. Math. **126**, 119-250 (1997)

25. —: Résultant anisotrope. Preprint.

26. Knudsen, F. and Mumford, D.: The projectivity of the moduli space of stable curves. I: Preliminaries on "det" and "Div". Math. Scand. **39**, 19-55 (1976)

27. König, J.: Einleitung in die allgemeine Theorie der algebraischen Größen. B. G. Teubner, Leipzig 1903

28. Koppenhöfer, D.: Determining the monogeneity of a quartic number field. Math. Nachr. **172**, 191-198 (1995)

29. Kraft, J.: Singularity of monomial curves in A^3 and Gorenstein monomial curves in A^4. Can. J. Math. **37**, 872-892 (1985)

30. Macaulay, F.S.: The algebraic theory of modular systems. Cambridge Univ. Press 1916 (Reprint, Cambridge Univ. Lib. 1994)

31. Matsumura, H.: Commutative ring theory. Cambridge stud. in adv. math. **8**. Cambridge University Press 1986

32. Patil, D.P.: Minimal sets of generators for the relation ideals of certain monomial curves. Manuscr. math. **80**, 239-248 (1993)

33. Nagata, M.: Local rings. Intersc. Publ., New York 1962

34. Renschuch, B.: Elementare und praktische Idealtheorie. Deutscher Verlag d. Wiss., Berlin 1976

35. Scheja, G.: Über die Bettizahlen lokaler Ringe. Math. Ann. **155**, 155-172 (1964)

36. Scheja, G., Scheja, O. and Storch, U.: On regular sequences of binomials. Manuscr. math. **98**, 115-132 (1999)

37. Scheja, G. and Storch, U.: Lokale Verzweigungstheorie. Schriftenreihe Math. Inst. Univ. Fribourg **5**, 1-150 (1974)

38. —, —: Über Spurfunktionen bei vollständigen Durchschnitten. J. reine u. angew. Math. **278/279**, 174-190 (1975)

39. —, —: Quasi-Frobenius-Algebren und lokal vollständige Durchschnitte. Manuscr. math. **19**, 75-104 (1976). Addendum **20**, 99-100 (1977)

40. —, —: Residuen bei vollständigen Durchschnitten. Math. Nachrichten **91**, 157-170 (1979)

41. —, —: On projective representations of finite algebras by graded complete intersections. Rend. Sem. Mat. Univ. Pol. Torino. **49**, 1-28 (1991)

42. —, —: On finite algebras projectively represented by graded complete intersections. Math. Z. **212**, 359-374 (1993)

43. —, —: Regular sequences and elimination. Commutative Algebra, Workshop Trieste 1992. Eds. A. Simis, N. V. Trung, G. Valla. World Sci. Publ. Singapore 217-236 (1994)

44. —, —: The divisor of the resultant. Beitr. zur Alg. u. Geom. – Contr. to Alg. and Geom. **37**, 149-159 (1996)

45. Serre, J.-P.: Algèbre Locale · Multiplicités. Lecture Notes in Math. **11**. Springer, Berlin ²1965.

46. Vasconcelos, W.V.: A note on normality and the module of differentials. Math. Z. **105**, 291-293 (1968)

47. Wall, C.T.C.: Weighted homogeneous complete intersections. Alg. geom. and sing. (La Rábida, 1991), Prog. Math. **134**, 277-300. Birkhäuser, Basel (1996)

48. Wiebe, H.: Über homologische Invarianten lokaler Ringe. Math. Ann. **179**, 257-274 (1969)

Index

admissible pair of weights and degrees 63, 74, 92, 122
—, strictly 68, 74, 90, 91, 92, 108, 113, 114, 118, 131, 135
admissible sequence of exponents 63, 74
—, strictly 68, 74, 78
Apéry basis of a num. monoid 16
arithmetic progression (of weights) 20
—, almost 32
Bezout domain 9
Bourbaki, Nicolas 55, 106
Brauer, Alfred 20
canonical class (loc. compl. inters.) 45, 48, 49, 51, 58
canonical module 14
Cartier divisor 126
Chardin, Marc 113
complete intersection (num. monoid) 21, 25, 28, 29ff., 38f., 92
complete intersection 46ff., 131
—, absolute 42, 50
—, finite 48, 54, 85, 101, 102, 123, 132
—, graded 53ff.
—, locally 42ff.
—, —, point criterion of 46
conormal module 43, 44ff., 51f.
content (of a polynomial) 1, 4
corner (binomial) 30, 32ff.
Davis, Edward D. 104
Davison, J.L. 32
Dedekind domain 9, 10, 49, 102, 103, 118
defining functions, globally 47f., 53f.
—, locally 47f., 53

degree formula 129ff.
degree of elimination 90, 92, 109, 118
—, principal 90
degree of singularity 13ff., 16, 34, 37f.
Delorme, Charles 36, 70
deviation of a local ring 25, 50
deviation of a num. monoid 25, 30, 38
distinguished sequence 70f.
divisor class group 10, 108
dualizing module 14, 35
—, reduced 133
—, relative 9, 48, 52
eliminant 123f.
elimination ideal 81ff., 91, 98ff., 104
—, generic 83, 84, 85, 89
elimination, main case 84, 85ff., 96ff.
elimination, main theorem 82
embedding dimension 50
fibre (algebra) 11, 42, 48
Frobenius algebra 9, 49
Frobenius number 14ff., 26, 29ff., 34, 37f., 39f., 114, 118
gaps of a num. monoid 13
Gauß, Carl Friedrich 11
generic binomials 69ff.
generic polynomials 60ff., 77
—, admissible sequence 63ff., 74, 80, 85ff.
—, —, strictly 68, 70f., 73, 74, 75, 78, 79, 80, 89, 90, 110, 131
genus of a num. monoid 13
Gorenstein ring (of dimension 1) 15, 102, 103, 111
Hauptkomponente 65
Herzog, Jürgen 31, 32

Hilbert series   15
Hurwitz, Adolf   82
ideal of a num. monoid   13ff.
—, canonical   14
—, dualizing   14, 34
inertia forms   82ff., 86f., 98f., 102,
      109f., 115
Johnson, S.M.   18, 31, 32
Jouanolou, Jean-Pierre   74
Kähler differentials   11, 51, 80
Knudsen, Finn   127
König, Julius (Gyula)   10, 11
Koszul complex   51, 52, 55ff., 106,
      112, 113, 119, 127, 130
Kraft, Jürgen   31, 34
Kronecker, Leopold   1, 10, 11
Kronecker extension   2ff., 6ff., 49, 56,
      77, 113, 116, 132
—, generalised   10f.
Kronecker's method of indeterminates
      1, 6
Krull domain   10
Krull, Wolfgang   29
Kummer, Ernst Eduard   10
Kummer, Renate   66
Macaulay, Francis Sowerby   113
Mumford, David   127
Nagata, Masayoshi   1, 3, 43, 63
norm formula   133f.
numerical monoid   13ff.
Patil, Dilip P.   32
peripheral component   87, 92
Picard group   8, 10, 48
Poincaré series   15, 19, 24, 27ff., 33,
      37, 53, 57, 91, 92, 114, 129ff.
Poisson formula   133
primitive polynomial   1
1-primitive polynomial   10
principal component   65, 75, 77, 79,
      80, 87, 90, 115
product formula   135f.

projective class group   45, 51, 58
Prüfer domain   9
quasi-Frobenius algebra   9
reduction formula   136
reduction in codimension 1   18ff., 21,
      39
reduction map   18ff., 39f.
regular sequence   41f., 43, 44, 53, 71,
      98f., 103, 104, 123
—, generic   61ff.
—, locally   42
—, strongly   41f., 53, 62, 64
relations of a num. monoid   23ff.
resultant   119ff., 129ff.
resultant ideal   107f., 109ff., 119, 121,
      128
sating numbers   16ff., 28, 29, 30f.,
      38ff., 114, 118
—, very   16ff., 30f., 38ff., 114, 118
saturated case   60, 63, 68, 74, 90ff.,
      114, 118, 122
Selmer, Ernst S.   20
separable algebra   11
socle determinant   96ff., 105f., 121f.
splitting of a num. monoid   36ff.
standard basis of a num. monoid   16
support of exponents   25, 61
Sylvester determinant   91, 113
Sylvester, James Joseph   20
symmetric num. monoid   15, 20, 21,
      29, 39
toric component   87
Trägheitsformen   82
type (set)   15, 34, 38
U-resultant   123f.
Vasconcelos, Wolmer V.   44, 51
Wall, Charles T.C.   63
Watanabe, Keiichi   36
weights (of indeterminates)   13, 60
Weil divisor   127
Wiebe, Hartmut   96, 101

Printed in the United States
by Baker & Taylor Publisher Services